Foundations of Electricity & Electronics

Foundations of Electricity & Electronics

Humphrey Kimathi
Ron. Bertrand

Disclaimer of Liability

The material contained in this publication are supplied without guarantee. The authors expressly disclaim all and any liability to any persons whatsoever in respect of anything done or omitted to be done by any such person in reliance either in whole or in part upon any contents of this publication.

Contents

Chapter 1 - Basic Electricity - I

ELECTRICITY

The complicated electronic systems involved in modern day communication, satellites, nuclear power plants, radio and television and even up to date automobiles, do not require technicians to understand the functioning of electric and electronic circuits. Modern day electronics is very modular. A remove and replace, or substitution of the suspected 'faulty module' is the approach to modern electronics servicing. This, in itself, is not a bad thing, as in the real world, getting an electronic device up and going is the most important thing. However, to have a true understanding requires a strong foundation in the basics of electricity. The term "electronic" infers circuits ranging from the first electronic device, the electron tube, to the newer solid state devices such as diodes and transistors, as well as integrated circuits (ICs). The term "electric" or "electrical" is usually applied to systems or circuits in which electrons flow through wires but which involve no vacuum tubes or solid state devices. Many modern electrical systems are now using electronic devices to control the electric current that flows in them.

What makes such a simple thing as an electric lamp glow? It is easy to pass the question off with the statement, "The switch connects the light to the power lines and it glows" or something to that effect. But what does connecting the light to the power lines do? How does energy travel through solid copper wires? What makes a motor turn? A radio play? What is behind the dial that allows you to pick out one radio station from thousands of others operating at the same time? How fast is electricity? There are no single, simple answers to any of these questions. Each question requires the understanding of many basic principles. By adding one basic idea to another, it is possible to answer, eventually, all of the questions that may be asked about the intriguing subjects of electricity, electronics and radio.

When a light switch is turned on and elsewhere the light suddenly glows, energy has found a path through the switch to the light. The path used is usually along copper wires and the tiny particles that do the moving and carry the energy are called electrons. These electrons are important to anyone studying electronics and radio since they are usually the only particles that are considered to move in electric circuits.

To explain what is meant by an electron, it will be necessary to investigate more closely the make-up of all matter. The word "matter" means, in a general sense, anything that can be touched. It includes substances such as rubber, salt, wood, water, glass, copper and air. The whole world is made of different kinds of stuff. The ancient Greek philosophers were always trying to find the 'stuff' that made up the universe. Even before the Greeks, the Alchemists were trying to find the basic building blocks that all matter was made from, though most of the time their driving force was not so much science but the pursuit of wealth. They figured that if they could isolate the building blocks of matter, then they would be able to 'create' matter themselves. One of their primary pursuits was the creation of the precious metal gold.

Water is one of the most common forms of 'stuff' that we call matter. If a drop of water is divided in two and then divided again and again until it can be divided no longer and still be water, then we have arrived at the smallest possible piece of wa-ter. We have a water molecule. The ancient Greeks would have called the smallest droplet of water an atomos (atom). The word atomos means indivisible. We know today that substances such as water can be divided into more fundamental bits.

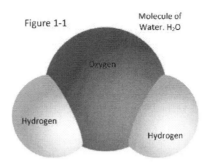

Figure 1-1

Molecule of Water. H_2O

Oxygen

Hydrogen

Hydrogen

The water molecule can be broken down into still smaller particles, but these new particles will not be water. Physicists have found that three smaller particles make up a molecule of water. Two atoms of hydrogen (H) and one atom of oxygen (O) are shown in figure 1-1. The symbols 'H' and 'O' are universal symbols used to represent Hydrogen and Oxygen. Oxygen, at average temperatures, is one of the several gases that constitute the air we breathe. Hydrogen is also a gas in its natural state; it is found in everyday use as part of the gas used for heating or cooking. If a gaseous mixture containing two parts of hydrogen and one part of oxygen is ignited, a violent chemical reaction, an explosion, will occur as water is formed and excess energy released. This is not an experiment that I would recommend.

Water is made up of two types of atoms, hydrogen and oxygen. Water is a molecule. A molecule is a substance that is made up of groups of atoms. If you divided a droplet of water down to its smallest possible size, you would have a single molecule of water. If you had the means to split the water molecule further, you would no longer have water; you will have the atoms (hydrogen and oxygen) that make up water. The chemical name of water then is Dihydrogen Oxide. It has been found that atoms are also divisible. An atom is made up of at least two types of particles: protons and electrons and a third particle called a neutron. Don't let these names concern you too much. For our purposes, the primary particle is the electron. Electrons and protons are called electrical particles and neither one is divisible (in typical environments). All the molecules that make up all matter in the universe are composed of these electrical proton-electron pairs.

ELECTRONS AND PROTONS

Electrons are the smallest and lightest of the fundamental particles. They are said to have a negative charge, meaning that they are surrounded by an invisible force field that will react in an electrically negative manner with other matter.

Protons are said to have a positive charge and are surrounded by an invisible force field that causes them to react in an electrically positive manner. The words negative and positive are just names to describe the so called charge of electrons and protons and their charge describes how they interact with each other. We could just as easily call the charge of the electron the white charge and the charge of the proton the black charge. My point is, 'charge' is an electric behaviour and since there are two types of charge we need to name them so that when we talk about them we will know which behaviour we are speaking of: either the positive charge behaviour or the negative charge behaviour.

5

AN EXPERIMENT WITH CHARGE

You may have already done this if you have, please try it again. Tear up some tiny strips of paper and place them on the table in front of you. Make sure no one is watching! Now run a hair comb through your hair briskly several times and put the comb close to the bits of paper. Before the comb touches the paper, a bit of paper will leap off the table, move through the air and cling to the comb. This happens because you have produced a charge on the comb, which will physically interact with matter around it (in our case the bits of paper).

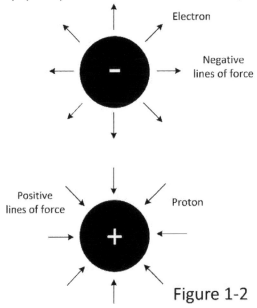

Figure 1-2

The charge on the comb was created by friction between the comb and your hair. Protons are about eighteen hundred times more massive than electrons and have a positive electric field surrounding them. The proton is exactly as positive as the electron is negative; each has a unit of electric charge.

When an electron and a proton are far apart, only a few of their lines of force (the invisible field around them) join and pull together. The attracting pull between the two charges is, therefore, small. When brought closer together, the electron and proton can link more of their lines of force and will pull together with greater force. If close enough, all the lines of force from the electron are joined to all the lines of force of the proton and there is no external field and they attract each other strongly. Together, a positive charged proton and the negative charge of an electron cancel out and they form a neutral, or uncharged, group. The neutral atomic particle, known as a neutron, exists in the nucleus of all atoms heavier than hydrogen. The fact that electrons repel other electrons, protons repel other protons, but electrons and protons attract each other gives us the basic law of charges:

LIKE CHARGES REPEL, UNLIKE CHARGES ATTRACT

Because the proton has about 1,800 times more mass than the electron, it seems reasonable to assume that when an electron and a proton attract each other, it will be the lower mass and free to move electrons that will do most of the actual moving. Such is the case. It is the electron which moves in electricity. If the proton was the lower mass particle and not in the nucleus, we would probably have called what we know today as 'electricity', something like 'proton'icity. Regardless of the difference in appar-ent size and mass, the negative field of an electron is just as strong negatively as the positive field of a proton is positive.

Though physically small, the field near the electron is quite strong. If the field strength (field strength = the strength of the invisible field) around an electron at a distance of one millionth of a metre is a certain amount, at two-millionths of a metre it will be one- quarter as much; at four millionths of a metre it will be one-sixteenth as much; and so on. If the field decreases as distance increases, the field is said to vary inversely with distance. It varies inversely with the distance squared.

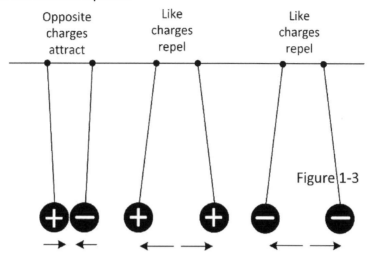

Figure 1-3

Note: A millionth of a metre has a name; it is called a 'micron'. A micron is not an *International System* (SI) Unit, but it is a unit which has stuck and is still often used. It is too nice of a name not to be used.

When an increase in something produces an increase in something else, the two things are said to vary directly rather than inversely. Two million electrons on an object produce twice as much negative charge as one million electrons would. The charge is directly proportional to the number of electrons. The invisible fields surrounding electrons and protons are known as electrostatic fields. The word 'static' means, in this case, *stationary*, or "not caused by movement", When electrons are made to move, the result is dynamic electricity. The word "dynamic" indicates that motion is involved. To produce a movement of an electron, it will be necessary to have either, a negatively charged field to push it, or positively charged field to pull it. Normally in an electric circuit, both a negative and a positive charge are used (a pushing and pulling pair of forces).

THE ATOM AND ITS FREE ELECTRONS

As of 2019 there are 118 different kinds of atoms, or elements, from which the millions of different forms of matter found in the universe, are composed. Let me elaborate on the last paragraph. About 100 different atoms are found in nature. Many atoms do not occur naturally. Many are only manufactured in very powerful particle colliders or inside stars.

The heavy atoms, those containing a large number of protons, electrons and neutrons, like uranium and radium are unstable. They throw off energy (they are radioactive) and decompose until they become stable non-radioactive atoms. A material, which is only made from one type of atom, is called an element. Water is not an element because it contains two types of atoms, hydrogen and oxygen. Water is, therefore, a molecule. Copper contains only copper atoms, so copper is an element. There are many other common elements.

The simplest and lightest atom (or element) is hydrogen. An atom of hydrogen consists of one electron and one proton, as shown in figure 1-4. In one respect the hydrogen atom is similar to all others: the electron whirls (orbits) around the proton, or nucleus, of the atom, much as planets rotate around the sun. Electrons whirling around the nucleus are termed planetary, or orbital, electrons! This is not an accurate model but is accurate for our purposes in radio physics. Electrons are more like clouds around the nucleus with electrons having a probability that they are more likely to be in the vicinity of one or more bands. We really can't say that they are just like planets around the sun, however we can use this analogy for our purposes.

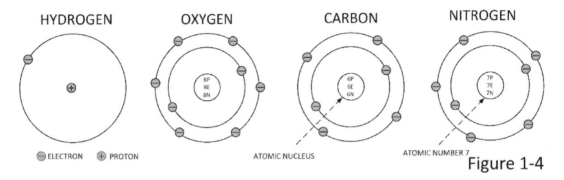

Figure 1-4

The nucleus is just a name given to the 'centre' of an atom. The next atom in terms of atomic weight is helium, having two protons and two electrons. The third atom is lithium, with three electrons and three protons and so on. All elements are grouped in families with the atomic weights in the Periodic table of elements.

Most atoms have a nucleus (centre) consisting of all the protons of the atom and also one or more neutrons. The electrons (always equal in number to the protons in the nucleus) are whirling around (orbiting) the nucleus in various layers. The first layer of electrons outside the nucleus can accommodate only two electrons. If the atom has three electrons, two will be in the first layer and the third will be in the next layer. The second layer is completely filled when eight electrons are whirling around in it. The third is filled when it has eighteen electrons. Some of the electrons in the outer orbit, or shell, of the atoms of many materials such as copper or silver, exist in a higher "conduction level" and can be dislodged easily. These electrons travel out into the wide open spaces between the atoms and molecules and may be termed free electrons.

Periodic Table of the Elements

1 H Hydrogen																	2 He Helium
3 Li Lithium	4 Be Beryllium											5 B Boron	6 C Carbon	7 N Nitrogen	8 O Oxygen	9 F Fluorine	10 Ne Neon
11 Na Sodium	12 Mg Magnesium											13 Al Aluminium	14 Si Silicon	15 P Phosphorus	16 S Sulphur	17 Cl Chlorine	18 Ar Argon
19 K Potassium	20 Ca Calcium	21 Sc Scandium	22 Ti Titanium	23 V Vandium	24 Cr Chromium	25 Mn Manganese	26 Fe Iron	27 Co Cobalt	28 Ni Nickel	29 Cu Copper	30 Zn Zinc	31 Ga Gallium	32 Ge Germanium	33 As Arsenic	34 Se Selenium	35 Br Bromine	36 Kr Krypton
37 Rb Rubidium	38 Sr Strontium	39 Y Yttrium	40 Zr Zirconium	41 Nb Niobium	42 Mo Molybdenum	43 Tc Technetium	44 Ru Ruthenium	45 Rh Rhodium	46 Pd Palladium	47 Ag Silver	48 Cd Cadmium	49 In Indium	50 Sn Tin	51 Sb Antimony	52 Te Tellurium	53 I Iodine	54 Xe Xenon
55 Cs Cesium	56 Ba Barium	57-71	72 Hf Hafnium	73 Ta Tantalum	74 W Tungsten	75 Re Rhenium	76 Os Osmium	77 Ir Iridium	78 Pt Platinum	79 Au Gold	80 Hg Mercury	81 Tl Thallium	82 Pb Lead	83 Bi Bismuth	84 Po Polonium	85 At Astatine	86 Rn Radon
87 Fr Francium	88 Ra Radium	89-103	104 Rf Rutherfordium	105 Db Dubnium	106 Sg Seaborgium	107 Bh Bohrium	108 Hs Hassium	109 Mt Meitnerium	110 Ds Darmstadtium	111 Rg Roentgenium	112 Cn Copernicium	113 Nh Nihonium	114 Fi Flerovium	115 Mc Moscovium	116 Lv Livermorium	117 Ts Tennessine	118 Og Oganesson

Lanthanide Series	57 La Lanthanum	58 Ce Cerium	59 Pr Praseodymium	60 Nd Neodymium	61 Pm Promethium	62 Sm Samarium	63 Eu Europium	64 Gd Gadolinium	65 Tb Terbium	66 Dy Dysprosium	67 Ho Holmium	68 Er Erbium	69 Tm Thulium	70 Yb Ytterbium	71 Lu Lutetium
Actinide Series	89 Ac Actinium	90 Th Thorium	91 Pa Protactinium	92 U Uranium	93 Np Neptunium	94 Pu Plutonium	95 Am Americium	96 Cm Curium	97 Bk Berkelium	98 Cf Californium	99 Es Einsteinium	100 Fm Fermium	101 Md Mendelevium	102 No Nobelium	103 Lr Lawrencium

Other electrons in the outer orbit will resist dislodgement and are called bound or valence electrons. Materials consisting of atoms (or molecules) having many free electrons will allow an easy interchange of their outer shell electrons while atoms with only bound electrons will hinder any electron exchange. Copper, for example, has one electron in its outer orbit or layer.

This lonely little outer electron of the copper atom is very easy to 'steal' from the copper atom and made to move. The outer electron is called a ***free electron.*** It is not free but loosely bound to the atom and easy to encourage away and made to move, so we call it a free electron. Copper does not resist the movement of its outer electrons strongly, or in other words, it does not offer much resistance to us if we try to get its outer electrons to move. We will talk about how we get them to move later. A material, which does not have free electrons, is said to have a high resistance. All metals have free electrons. Most common metals when heated cause greater energy to be developed in their free electrons. The more energy electrons have the greater they resist orderly movement through the material. The material is said to have an increased resistance to the movement of electrons through it.

THE ELECTROSCOPE

The charge on the comb was created by friction between the comb and your hair. Protons are about eighteen hundred times more massive than electrons and have a positive electric field surrounding them.

An example of electrons and electric charges acting on one another is demonstrated by the action of an electroscope. An electroscope consists of two very thin gold or aluminium leaves attached to the bottom of a metal rod. The delicate metal foil leaves and rod are encased in a glass bottle to protect them from air movement.

To understand the operation of the electroscope, it is necessary to recall these facts:

1) Normally an object has a neutral charge.
2) Like charges repel; unlike charges attract.
3) Electrons are negative.
4) Metals have free electrons.

Normally the metal rod of the electroscope has a neutral charge and the leaves hang downward parallel to each other, as shown in figure 1-5A. The leaves are shown in charged position in figure 1-5B and 1-5C. With no charge the leaves hang vertically down from the rod.

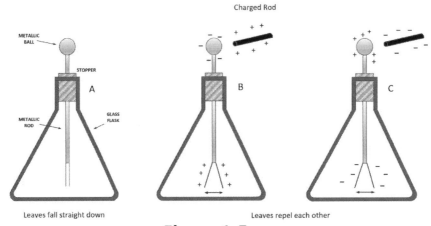

Figure 1-5

Rubbing a piece of hard rubber with wool causes the wool to lose electrons to the rubber and the excess electrons on the rubber, charge the rubber negatively. When such a negatively charged object is brought near the top of the rod, some of the free electrons at the top are repelled and travel down the rod, away from the negatively charged object. Some of these electrons force themselves onto one of the leaves and some onto the other. Now the two leaves are no longer neutral but are slightly negative and repel each other, moving outward from the vertical position as shown. When the charged object is removed, electrons return up the rod to their original areas. The leaves again have a neutral charge and hang down parallel to each other.

Since the charged object did not touch the electroscope, it neither placed electrons on the rod nor took electrons from it. When electrons were driven to the bottom, making the leaves negative, these same electrons leaving the top of the rod left the top positive. The overall charge of the rod remained neutral. When the charged object was withdrawn, the positive charge at the top of the rod pulled the displaced electrons up to it. All parts of the rod were then neutral again. If a positively charged object, such as a glass rod vigorously rubbed with a piece of silk, is brought near the top of the electroscope rod, some of the free electrons in the leaves and rod will be attracted upward toward the positive object. This charges the top of the rod negatively because of the excess of free electrons there. Both leaves are left with a deficiency of free electrons which means they are positively charged. Since both leaves are similarly charged again, they repel each other and move outward a second time.

A deficiency of electrons on an object leaves the object with a positive charge. An excess of electrons gives it a negative charge. If a negatively charged object is touched to the metal rod, some excess electrons will be deposited onto the rod and will be immediately distributed throughout the electroscope. The leaves spread apart. When the object is taken away, an excess of electrons remains on the rod and the leaves. The leaves stay spread apart. If the negatively charged electroscope is touched to a large body that can accept the excess free electrons, such as a person, a large metal object, or earth (the ground), the excess electrons will have a path to leave the electroscope and the leaves will collapse as the charge returns to neutral. The electroscope has been discharged. If a positively charged object is touched to the top of the metal rod, the rod will lose electrons to it and the leaves will separate. When the object is taken away, the rod and leaves still lack free electrons and are therefore positively charged and the leaves will remain apart.

A large neutral body touched to the rod will drain some of its free electrons to the electroscope discharging it and the leaves will hang down once more. The electroscope demonstrates the free movement of electrons that can take place through metallic objects or conductors when electric pressures, or charges, are exerted on the free electrons.

11

2 - Basic Electricity - II

THE THREE BIG NAMES IN ELECTRICTY

Without calling them by name, we have touched on the three elements always present in operating electric circuits:

Current: A progressive movement of free electrons along a wire or other conductor caused by electrical pressure.

Voltage: The electron moving force in a circuit that pushes and pulls electrons (current) through the circuit. Also called electromotive force and electrical pressure.

Resistance: Any opposing effect that hinders free electron progress through wires when an electromotive force is attempting to produce a current in the circuit.

We will be talking a lot about these three properties of an electric circuit and how they relate to each other.

A SIMPLE CIRCUIT

The simplest of electric circuits consists of some electron moving force, or source, such as that provided by a dry cell, or battery, and a load, such as an electric light, connect-ing wires and a control device. A pictorial representation and the electric diagram of a simple circuit is shown in figure 2-1.

The diagram in figure 2-1 is called a schematic diagram and is much easier to draw. The control device is a switch, to turn the bulb ON and OFF. In effect, the switch disconnects one of the wires from the cell. In our circuit, the switch could be connected anywhere to turn the bulb ON and OFF. In this circuit, the light bulb is the load. Although the wires connecting the source of electromotive force (the dry cell) to the load may have some resistance, it is usually tiny in comparison with the resistance of the load and is ignored in most cases.

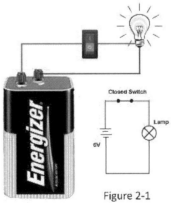

Figure 2-1

A straight line in a schematic diagram is considered to connect components electrically and does not represent any resistance in the circuit. In the simple circuit shown, the cell produces the electromotive force (voltage) that continually pulls electrons to its positive terminal from the bulb's filament and pushes them out of the negative terminal to replace the electrons that were lost to the load by the pull of the positive terminal. The result is a continual flow of electrons through the lamp filament, connecting wires and source. The special resistance wire of the lamp filament heats when a current of electrons flows through it. If enough current flows, the wire becomes white hot and the bulb glows and gives off light (incandescence).

CURRENT

A stream of electrons forced into motion by an electromotive force is known as a current. Here we have a definition of electric current:

Current is the ordered movement of electrons in a circuit

In a good conducting material such as copper, one or more free electrons in the outer ring are constantly flying off at a high rate. Electrons from other nearby atoms fill in the gaps. There is a constant aimless movement of billions of electrons in all directions at all times in every part of any conductor. This random movement of electrons is not an electric current, as there is not yet movement in any one direction.

Only when a voltage is applied, do we get an ordered movement of electrons. When an electric force is applied across the conductor (from a battery), it drives some of these aimlessly moving free electrons away from the negative force toward the positive. It is unlikely that any one electron will move more than a fraction of a centimetre in a second, but an energy flow takes place along the conductor at a significant fraction of the velocity of light.

Notice that I said the energy flow in the circuit is very fast (almost the speed of light - but not quite). The speed of the electrons in a circuit - or the current flow is in fact very slow. I won't bore you with calculations. However, I did once calculate how fast the electron flow was in a typical circuit and it came to be about walking speed. Electron flow or current flow is very slow. The effect of an electric current at a distance through a conductor, on the other hand, is very fast.

If you have trouble with this and many do, think about how fast water travels through a pipe. The dam where the water comes from may be many kilometres away from the tap. When you turn the tap on the water comes out immediately does it not? Did the water travel all the way from the dam to the tap in an instant? I am sure you would agree that it did not. If I tried to tell you that it did you would most probably laugh at me and say the water was already in the pipe, all you did by turning the tap on was to make the water move in the pipe between the dam and the tap.

Similarly, the electrons are already in the wire (conductor). When we close a switch in the circuit and apply an electromotive force, all we are doing is making all the electrons move in the conductor at the same time. It may take a very long time for an electron leaving the source to reach the load if it ever does.

HOW FAST ARE MARBLES?

Let's do another analogy to make sure we have got this clear. Suppose a pipe was connected between Sydney and Melbourne. Imagine if we blocked off one end of the pipe and filled it with marbles until we could not fit any more in. We now unblock the pipe and we have a crowd of people at each end to witness the experiment and find out how fast marbles travel. The two crowds are in contact by telephone or radio and anxiously awaiting the big moment. One too many marbles is about to be inserted into the full pipe. As soon as a marble is inserted in Sydney, a marble drops out in Melbourne. The effect of pushing a marble into the pipe in Sydney caused an immediate result or effect in Melbourne. The newspaper's report "Eccentric experi-menter proves that marbles travel at the speed of light". Is this right?

I hope you are shaking your head and saying 'no'. The marble, which fell out of the pipe in Melbourne, was sitting there ready to fall out as soon as the marble was pushed in at the Sydney end. So, marbles definitely do not travel at the speed of light any more than do electrons in a conductor. The marbles were already in the pipe just as the electrons were already in the conductor.

The effect of an electric current at a distance is almost instantaneous; however, the speed of the electrons is very slow. A source of electrical energy does not increase the number of free electrons in a circuit; it merely produces a concerted pressure on loose, aimlessly moving electrons. If the material of the circuit is made of atoms or molecules that have no freely interchanging electrons, the source cannot produce any current in the material. Such a material is known as an insulator, or a non-conductor.

The amount of current in a circuit is measured in Amperes, abbreviated 'A' or Amp. An ampere is a certain number of electrons passing or drifting passed a single point in an electric circuit in one second. Therefore, an ampere is a rate of flow, similar to litres (or marbles) per minute in a pipe.

The quantity of electrons used in determining an ampere (and other electrical units) is the Coulomb, abbreviated 'C'. An ampere is one Coulomb per second. A single coulomb is 6,250,000,000,000,000,000 electrons. This large number is more easily expressed as 6.25×10^{18}, which is read verbally as 6 point 25 times by 10 to the eighteenth power".

ELECTROMOTIVE FORCE OR VOLTAGE

The electron moving force in electricity, variously termed electromotive force (emf), electric potential, potential difference (PD), electric pressure and voltage (V), is responsible for the pulling and pushing of the electric current through a circuit. The force is the result of an expenditure of some form of energy to produce an electrostatic field.

An emf (I like to read this as 'electron-moving force') exists between two objects whenever one of them has an excess of free electrons and the other has a deficiency of free electrons. An object with an excess of electrons is negatively charged. Similarly, an object with a deficiency of electrons is positively charged. Should two objects with a difference in charge be connected by a conductor, a discharge current will flow from the negative body to the positive one.

Instead of writing negatively or positively charged all the time it is easier to say **'-Ve'** or **'+Ve.'** I tend not to use this notation, but you should know what it means.

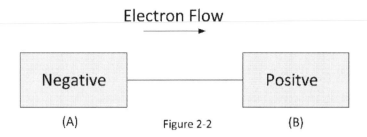

Figure 2-2

(A)   (B)

An emf also exists between two objects whenever there is a difference in the number of free electrons per unit volume of the objects. In other words, both objects may have a negative charge, but one is more negative than the other. The less negative object is said to be positive with respect to the more negative object.

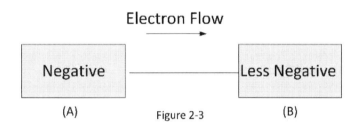

Figure 2-3

(A)   (B)

In the electrical trades, it is common to hear of current flow from positive to negative. This is called the **conventional** direction of current flow. This is just what it says, a con-vention (popular method). Current flow is electron flow and it is from negative to posit-ive. Conventional current flow is mostly used in Electrical Engineering. This book and that is to use **electron flow;** current flows from negative to positive.

The unit of measurement of electric pressure, or emf, is the Volt (V). A single torch dry cell produces about 1.5V. A volt can also be defined as the pressure required to force a current of one ampere through a resistance of one ohm.

PRODUCING AN ELECTRON-MOVING FORCE

NOTE: A battery is a collection of cells (like a battery of cannons). There is no such thing as an AA battery; it is an AA cell. On the other hand, a 9 volt transistor battery or a car battery are samples of real batteries because they are constructed from a number of cells connected together (in series).

1) Chemical (cells and batteries).
2) Electromagnetic (generators).
3) Thermal (heating the junction of dissimilar metals).
4) Magnetostriction (filters and special energy converters called transducers).
5) Static (laboratory static electricity generators).
6) Photoelectric (light sensitive cells).
7) Magnetohydrodynamics (MHD, a process that converts hot gas directly to electric current).
8) Piezoelectricity - some materials produce a voltage when physical pressure is applied to them.

EFFECTS OF AN ELECTRIC CURRENT

1)   Heat and light - current flowing in a conductor causes the conductor's temperature to increase. If the temperature increases sufficiently, the conductor will become incandescent and radiate light.

2)   Magnetic - a conductor carrying a current will produce a magnetic field around the conductor.

3)   Chemical - electroplating, charging batteries. An electric current is able to cause a chemical reaction.

THE BATTERY IN A CIRCUIT

In the explanations thus far, "objects," either positively or negatively charged, have been used. A common method of producing an emf is by the chemical action in a battery. Without going into the chemical reactions that take place inside a cell, a brief outline of the operation of a Leclanche cell is given here.

Consider a torch battery. Such a battery (two or more cells form a battery) is composed of a zinc container, a carbon rod down the middle of the cell and a black, damp, paste -like electrolyte between them. The zinc container is the negative terminal. The carbon rod is the positive terminal. The active chemicals in such a cell are the zinc and the electrolyte.

The materials in the cell are selected substances that permit electrons to be pulled from the outer orbits of the molecules or atoms of the carbon terminal, chemically by the electrolyte, and be deposited onto the zinc can. This leaves the carbon positively charged and the zinc negatively charged.

The number of electrons that move is dependent upon the types of chemicals used and the relative areas of the zinc and carbon electrodes. If the cell is not connected to an electric circuit, the chemicals can pull a certain number of electrons from the rod over to the zinc. The massing of these electrons on the zinc produces a backward pressure of electrons, or an electric strain, equal to the chemical energy of the cell and no more electrons can move across the electrolyte.

If a wire is connected between the positive and negative terminals of the cell, the 1.5 V of emf starts a current of electrons flowing through the wire. The electrons flowing through the wire start to fill up the deficient outer orbits of the molecules of the positive rod.

The electron movement away from the zinc into the wire begins to neutralise the charge of the cell. The electron pressure built up on the zinc, which held the chemical action in check, is decreased. The chemicals of the electrolyte can now force an electron stream from the positive rod through the cell to the zinc, maintaining a current of electrons through the wire and battery as long as the chemicals hold out.

As soon as the wire begins to carry electrons; the electrolyte also has an electric current moving through it. This motion produces an equal amount of current through the whole circuit at the same time. This is a very important concept to understand. There are no bunches of electrons moving around an electric circuit like a group of racehorses running around a track. A closed circuit is more like the racetrack with a single lane of cars, bumper to bumper. Either all must move at the same time, or none can move. In an electric circuit, when electrons start flowing in one part, all parts of the circuit can be considered to have the same value of current flowing in them instantly. Most circuits are so short that the energy flow velocity, 300,000,000 meters per second, may be dis-regarded. Figure 2-4 shows the internal construction of a typical 12V lead acid cell and a 9V alkaline battery. Notice the individual cells that make up the battery. Six cells for the lead-acid battery and the 9V battery.

Figure 2-4

IONISATION

When an atom loses an electron, it lacks a negative charge and is therefore positive. An atom with a deficiency of one or more electrons is called a positive ion. On the other hand, if an atom were to gain an electron, albeit temporarily, it is a negative ion.

In most metals the atoms are constantly losing and regaining free electrons. They may be thought of as constantly undergoing ionisation. Because of this, metals are usually good electrical conductors. Atoms in a gas are not normally ionised to any great extent and therefore, a gas is not a good conductor under low electric pressures.

However, if the emf is increased across an area in which gas atoms are present, some of the outer orbiting electrons of the gas atoms will be attracted to the positive terminal of the source of emf and the remainder of the atom will be attracted toward the negative. When pressure increases enough, one or more free electrons may be torn from the atoms. The atoms are ionised. If ionisation happens to enough of the atoms in the gas, a current flows through the gas. For any particular gas at any particular pressure, there is a certain voltage value that will produce ionisation. Below this value, the number of ionised atoms is small. Above the critical value, more atoms are ionised, producing greater current flow, which tends to hold the voltage across the gas at a constant value. In an ionised condition the gas acts as an electric conductor. Examples of ionisation of gases are lightning, neon lights and fluorescent lights. Ionisation plays an important part in electronics and radio.

TYPES OF CURRENT AND VOLTAGE

Different types of currents and voltages that are dealt with in electricity:
1. Direct current (DC). There is no variation of the amplitude (strength) of the current or voltage. Supplied from batteries, DC generators and power supplies.
2. Varying direct current (VDC). The amplitude of the current or voltage varies but never falls to zero. Found in many radio and electronic circuits. A telephone is a good example of the use of varying direct current.
3. Pulsating direct current (PDC). The amplitude drops to zero periodically (such as our light bulb circuit if it was repeatedly switched on and off).
4. Alternating current (AC). Electron flow reverses (alternates) periodically and usually changes amplitude in a more or less regular manner.

AC is produced in AC generators, oscillators, some microphones and radios in general. Household electricity is alternating current.

17

RESISTANCE

Resistance is that property of an electric circuit which opposes the flow of current. Resistance is measured in Ohms. The higher the resistance in an electric circuit the lower the current flow. The symbol used for resistance is the Greek letter Omega - Ω. If a circuit with an electric pressure of 1 volt causes a current of 1 Ampere to flow, then the circuit has a Resistance of 1Ω.

What is resistance? It is all well and good to say it is the opposition to current flow but from where does the opposition come? Resistance causes heat. In fact, resistance is the only electrical property that produces heat. Resistance is the effort or energy that it takes to get electrons moving. When we do get them moving some of the energy is dissipated as heat. It is lost energy. So, resistance is the energy lost making electrons move in an orderly manner.

WHATS IN A NAME

We have learned quite a few new terms. Some of these terms are taken from people's names. These people were usually pioneers in the fields of physics, electricity or electronics. Read the very short biographies below and think about the person's name and what it represents in an electric circuit.

George Simon Ohm. Born March 16, 1789, Erlangen, Bavaria [Germany]. Died July 6, 1854, Munich. German physicist who discovered the law named after him, which states that the current flow through a conductor is directly proportional to the potential difference (voltage) and inversely proportional to the resistance.

Andre-Marie Ampere. Born Jan. 22, 1775, Lyon, France. Died June 10, 1836, Marseille. French physicist who founded and named the science of electrodynamics, now known as electromagnetism. Ampere was a prodigy who mastered all mathematics then existing by the time he was 12 years old. He became a professor of physics and chemistry at Bourg in 1801 and a professor of mathematics at the Ecole Polytechnique in Paris in 1809.

Allesandro Giuseppe Antonio Anastasia Volta. Born Feb. 18, 1745, Como, Lombardy [Italy]. Died March 5, 1827, Como. Italian physicist whose invention of the electric battery provided the first source of continuous current. He became a professor of physics at the Royal School of Como in 1774 and discovered and isolated methane gas in 1778. One year later he was appointed to the chair of physics at the University of Pavia.

Charles Augustin de Coulomb. Born June 14, 1736, Angouleme, France. Died Aug. 23, 1806, Paris. French physicist best known for the formulation of Coulomb's law, which states that the force between two electrical charges is proportional to the product of the charges and inversely proportional to the square of the distance between them.

Goerges Leclanche. Born 1839, Paris. Died Sept. 14, 1882, Paris. French engineer who in about 1866 invented the battery bearing his name. In slightly modified form, the Leclanche battery, now called a dry cell, is produced in great quantities and is widely used in devices such as torches and portable radios.

3 - Basic Electricity - III

MORE ON RESISTANCE

As discussed briefly in Basic Electricity Part II, resistance is the opposition to current flow in any circuit. Resistance is measured in ohms and the symbol for resistance is Ω, though for equations (we will be using them soon) the letter 'R' is used.

Copper and silver are excellent conductors of electric current. When the same emf (voltage) is applied across an iron wire of equivalent size compared to silver, only about one-sixth as much current flows. Iron is considered to be good conductor.

When the same voltage (emf) is applied across a length of rubber or glass, no electron drift results. These materials are insulators. Insulators are used between conductors when it is desired to prevent electric current from flowing between them. To be more precise, in a normal 'electric' circuit the current flow is negligible through an insulator.

Silver is one of the better conductors and glass is one of the best insulators. Between these two extremes are found many materials of intermediate conducting ability. While such materials can be catalogued as to their conducting ability, it is more usual to think of them by their resisting ability. Glass (when cold) completely resists the flow of current. Iron resists much less. Silver has the least resistance to current flow.

The resistance of a wire or other conducting material is dependent on four physical factors:

1. The type of material from which it is made (silver, iron, etc.).
2. The length (the longer the conductor, the greater the resistance).
3. The cross-sectional area of the conductor (larger area, more free electrons and less resistance).
4. Temperature (the higher the temperature, the greater the resistance, except for carbon and other semiconductor materials).

A piece of silver wire of given dimensions will have less resistance than an iron wire of the same dimensions. It is reasonable to assume that if a 1-metre piece of wire has a 1 ohm resistance, then 2 metres of the same wire will have 2 ohms of resistance.

On the other hand, if a 1 metre piece of wire has 1 ohm of resistance then two pieces of this wire placed side by side will offer twice the cross-sectional area, will conduct current twice as well and therefore will have half as much resistance.

$$R = \rho L / A$$

Where:

R = resistance in ohms.

ρ = resistivity of the material in ohms per metre cube.

L = length in metres.

A = cross-sectional area of the wire in square metres.

ρ is a Greek letter spelt 'rho' and pronounced the same as in 'row-your- boat.'

RESISTIVITY ρ

Since the resistance of a wire (or any other material) depends on its shape, we must have a standard shape to compare the conducting properties of different materials. This standard is a cube measuring 1 metre on each side. That's a pretty big cube! Smaller material sizes are measured and then the results extrapolated to an Ohm-metre. The resistance measured between opposite faces of the cube is called the resistivity.

Resistivity should not be confused with resistance. The resistivity of a material is the resistance measured for a standard size cube of that material. If you look at the table below you will see for example that the resistivity of copper is 1.76×10^{-8}• The part of this number shown as 10^{-8} is called the exponent and the 'minus 8' means that the decimal point must be moved 8 places to the left. If we take 1.76 and move the decimal point 8 places to the left, we get:

0.000 000 017 6 ohm-metre (Ω.m)

Now this is a very low resistance indeed. It is the resistance that would be measured across the opposite faces of a cubic metre of solid copper. Resistivity is not resistance. Resistivity is a measure of the resistance of a material of a standard size to allow us to compare how well that material conducts or resists current compared to other materials.

The Resistivity of Metals and Alloys at 20 degrees C.

Material	Resistivity
Silver	1.62×10^{-8}
Copper	1.76×10^{-8}
Aluminium	2.83×10^{-8}
Gold	2.44×10^{-8}
Brass	3.90×10^{-8}
Iron	9.40×10^{-8}
Nickel	7.24×10^{-8}
Tungsten	5.48×10^{-8}
Manganin	45.0×10^{-8}
Nichrome	108×10^{-8}

The most common conducting material used in electrical circuits is, of course, copper, for it is a good conductor and relatively cheap. You can see in the table that aluminium is not as good a conductor as copper. However, aluminium is used for conductors more than any other material because of its light weight and corrosion resistance. In overhead power line distribution, weight is a crucial consideration, so aluminium is the conductor of choice. In radio and communications, antennas are made of aluminium, because of the light weight and ability to withstand corrosion.

CALCULATING THE RESISTANCE OF A WIRE

A 100 metre length of copper wire is used to wind the primary of a transformer and the wire has a diameter of 0.5 millimetres. What is the resistance of the winding?

Solution:

We need to use the equation $R = \rho l/A$

Since the wire is copper, we look at the Resistivity table and get a ρ of 1.76×10^{-8} for copper. The length (L) = 100 metres. We need to calculate 'A' (the cross-sectional area) from the equation for the area of a circle.

$A = \pi d^2/4$ or πr^2

Where:

π = Another Greek letter, the mathematical constant Pi, approximated by 22/7
d = diameter of the circle (wire).

$A = 22/7 \times (0.5 \times 10^{-3})^2 / 4$
$A = 3.142 \times 0.00000025 / 4$
$A = 0.000000196375$ square metres

$R = \rho l/A$
$R = 1.76 \times 10^{-8} \times 100 / 0.000000196375$
$R = 8.96$ ohms

$R = \rho L/A$ ← What does this say?

We shall look at the numerator and the denominator on the right-hand side separately.

$R = \rho L$

This means that resistance is directly proportional to the resistivity and to the length. As either p or L changes so does R. If p or L increases by say a factor of 2 then so does R. In other words, doubling the length of a wire doubles its resistance. If we increase the length by 3.25 times the resistance is increased 3.25 times. So, from the equation, we say R is directly proportional to length and resistivity.

The cross-sectional area is in the denominator of the equation on the right-hand side. Ignoring the numerator for the moment, we can rewrite this relationship as:
$R = 1/A$

This means that resistance is inversely proportional to the cross-sectional area, A. If the cross-sectional area of a wire was to be doubled then its resistance would be halved. This would be the same as twisting two wires together and using them as one. If the cross-sectional area of a wire was increased by a factor of say 4.5 times, then the resistance would be R/4.5 of what it originally was.

We have avoided bringing temperature into our calculations. The resistance of metals increases with temperature - we will discuss this further later.

In the real world, conductors are not round. Wire is not truly round. Copper circuit board track is definitely not round. In these cases, we work with the cross-sectional area given by the manufacturer. I would like you to know what this equation 'says' to you about the resistance of a wire.

Resistance is directly proportional to the length and resistivity and inversely proportional to the cross-sectional area.

THE METRIC SYSTEM

Scientific measurements are mostly done using the metric system. The metric system uses multiples of base 10. The basic prefixes of metric units of measurement are:

Prefix	Symbol	Meaning	Factor
Atto	(a)	quintillionth of	10^{-18} times
Femto	(f)	quadrillionth of	10^{-15} times
Pico	(p)	trillionth of	10^{-12} times
Nano	(n)	billionth of	10^{-9} times
Micro	(μ)	millionth of	10^{-6} times
Milli	(m)	thousandth of	10^{-3} times
Centi	(c)	hundredth of	10^{-2} times
Deci	(d)	tenth of	10^{-1} times
Unity		1	
Deka	(da)	ten times	10^{1} times
Hecto	(h)	hundred times	10^{2} times
Kilo	(k)	thousand times	10^{3} times
Mega	(M)	million times	10^{6} times
Giga	(G)	billion times	10^{9} times
Tera	(T)	trillion times	10^{12} times

The common prefixes used in electronics that you need to learn are underlined. You should need to memorise all of these. Volts, amperes, ohms, etc. may use metric based prefixes. Some examples of the utilisation of these prefixes are:

1 kV = 1000 volts.

1mV = one thousandth of a volt.

10MΩ = 10 million ohms

56mΩ = 56 thousandths of an ohm.

25mA = 25 thousandths of an ampere.

65μA = 65 millionths of an ampere.

CONVERTING FROM ONE PREFIX TO ANOTHER

Hertz is the UNIT. Mega and Kilo are prefixes of the unit.

Let us try the one above.

Megahertz is 106 - 106 means 1,000,000 or one million.
Kilohertz is 103 - 103 means 1,000 or one thousand.

Megahertz is a larger unit than kilohertz. How much larger is a megahertz than a kilohertz? How many times will one thousand divide into one million?

It takes 1,000 kilohertz to make one megahertz. A megahertz is 1000 times larger than a kilohertz, so to convert megahertz to kilohertz, we must multiply megahertz by 1000.

144MHz is the same as 144,000 kilohertz.

If you were asked to do it the other way round, that is, convert kilohertz to megahertz you would DIVIDE by 1000.

100kHz is 0.1MHz.

I think this looks all too easy to some and for this reason, it is done too quickly and very often the wrong answer is picked in the exam and 2 marks are lost all too easily.
Let's do another.

How do you convert microfarads to picofarads?

Please do it step by step, not what you immediately think it should be, unless you are very confident.
This would be my reasoning.

Micro= 1/1,000,000 or one millionth.
Pico = 1/1,000,000,000,000 or one millionth of a millionth.

This means that picofarads are one million times SMALLER than microfarads. To convert microfarads to picofarads MULTIPLY by 1,000,000.

Example - convert 0.001 microfarads to picofarads.

If we multiply 0.001 by 1,000,000, we get 1000 picofarads.

The easiest way to multiply by 1,000,000 is to move the decimal point 6 places to the right. Can you see that moving the decimal point in 0.001 six places to the right gives you 1000? Do it on a piece of paper - it's the way I do it. Simply put your pen on the decimal point in 0.001 and draw 6 'hoops' moving to the right. Then add "zeros" to the 'hoops' that have nothing in them.

TIP: Learn to use the ENG key on your scientific calculator. If you need to do a conversion then just enter the number you are given in scientific notation and then repeatedly press the ENG until you get the exponent you want. If the numbers go the wrong way press SHIFT ENG to reverse the direction. ENG and SHIFT ENG never change the number you have entered they just manipulate the exponent.

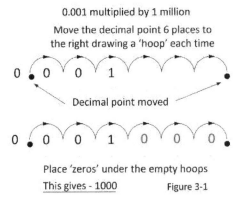

0.001 multiplied by 1 million

Move the decimal point 6 places to the right drawing a 'hoop' each time

Decimal point moved

Place 'zeros' under the empty hoops

This gives - 1000   Figure 3-1

This may seem like a silly way of doing it, but it is a safe way. If you had to divide by 1 million (converting picofarads to microfarads), use the same method only move the decimal point 6 places to the left.

You can use pen and paper in the exam. Using this method, you can convert any prefix to another prefix. Do not try and memorise the conversions, but do memorise what each prefix means, then work out how to convert one to the other using the techniques described above.

SYMBOLS IN TEXT

Since some assignments (when this book is used as part a course) are done using plain text you cannot show superscripts, subscripts and symbols. Many email programs do allow this, but you should avoid using symbols as when your email is received the symbols could be stripped out.

This is the method I use (but you can use your own as long as it is clear).

For example, show:

10^6 (10 to the power 6) as 10^6
10^{-6} (10 to the power minus 6) as 10^-6

Use the '^' which is a 'shifted 6' to indicate that what follows is a superscript. For symbols, you can type the symbol name in brackets. Take the formula for the resistance of a conductor that we used earlier in this chapter.

$R = \rho L/A$

The Greek letter ρ can be typed as (rho): R = (rho)L/A

If you have to type the Greek letter π in a formula you can just type (Pi).

THE RESISTOR COLOUR CODE

Resistors can be very small electronic components. Too small to write the resistor's value on, so instead, each resistor has colour coded bands which tell you its value and tolerance. The tolerance is the percentage of error, about which the resistor may vary from its coded value.

Resistors are manufactured in what are called preferred values. Usually, you will require a certain value of resistance in ohms and normally you will choose a resistance with the closest preferred value. If you want a specific resistance that does not match any preferred value, then you may have to make up a resistor especially for the job, or more commonly, you will use a variable resistance and adjust it using an ohmmeter.

There are only so many numerical values in a decade, i.e. from 0-10, or 0-100, 0-1,000, etc. Simple resistors only have 12 values in a decade e.g. 1.0, 1.2, 1.5, 1.8, 2.2, 2. 7, 3.3, 3.9, 4.7, 5.6, 6.8, 8.2. This is called the E12 series

A newer series called E24 has 24 standard values and closer tolerances.

E12 Series	E24 Series	
10	10	33
12	11	36
15	12	39
18	13	43
22	15	47
27	16	51
33	18	56
39	20	62
47	22	68
56	24	72
68	27	82
82	30	91

Resistors can have either 4 or 5 coloured bands. With each type, the last band is the tolerance. Five band resistors simply have room for three significant figures even though the E24 series only really needs two bands.

Most of the time these resistors are 1 or 2% tolerance (within +/- 1 or +/- 2% of the stated resistance). This will be either a brown or red band, respectively, at one end of the resistor, separated from the other bands. The value of the resistor starts at the other end.

The colour code, which you must commit to memory is:

Black	= 0	Green	= 5
Brown	= 1	Blue	= 6
Red	= 2	Violet	= 7
Orange	= 3	Grey	= 8
Yellow	= 4	White	= 9

The tolerance band of a resistor is coded:

Gold	= 5%	Silver = 10%
Brown	= 1%	Red = 2%

The tolerance band is always a band on the end and separated slightly from the other bands.

The second last band is the multiplier for both 4 and 5 colour banded resistors. The first letter of each word in the following sentence may help you to remember the colour code.

Big Boys Race Our Young Girls But Violet Generally Wins.

Example 1:

A resistor has 4 coloured bands. From left to right the bands are coloured:
Yellow, Violet, Yellow and Gold
The first significant digit is: 4.
The second significant digit is: 7.
The third band is the multiplier, in this case, 4 and means add four zeros.
So we get 470000Ω or 470kΩ.
The tolerance is (gold) +/- 5%.
So the final result is 470kΩ +/-5%.

Example 2:

A resistor has 5 coloured bands. From right to left the bands are coloured:

Green, Blue, Black, Red, Brown

The first significant digit is: 5.
The second significant digit is: 6.
The third significant digit is: 0.
The multiplier in the fourth band is 2 (add two more zeros).
So we get 56000 Ohms or 56kΩ.
The tolerance band is 1, therefore, 1% tolerance.
So the final result is 56kΩ +/- 1%
The fourth band on a five-banded resistor can be Gold or Silver. Gold means the multiplier is 0.1 and Silver means 0.01. To multiply by 0.1, move the decimal place one place to the left. To multiply by 0.01, move the decimal place two places to the left.

SURFACE MOUNT RESISTORS

Surface mounted resistors (SMD) are so small the resistor colour code is impractical. These resistors have a 3 or 4 character code to convey their resistance

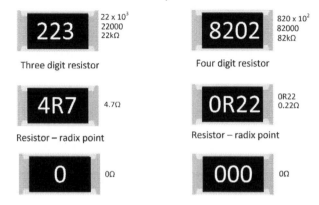

223	22×10^3 22000 22kΩ	
Three digit resistor		
8202	820×10^2 82000 82kΩ	
Four digit resistor		
4R7	4.7Ω	
Resistor – radix point		
0R22	0R22 0.22Ω	
Resistor – radix point		
0	0Ω	
000	0Ω	

CONDUCTANCE

There is no need for you to be concerned about the term 'conductance'. I mention it only for your information and you will occasionally see the term.

For some (not myself) it is more convenient to work in terms of the ease with which a current can be made to flow, rather than the opposition to current flow (resistance). Conductance is merely the reciprocal of resistance. The symbol for conductance is 'G' and G=l/R. Conductance is measured in Siemens 'S', formerly the 'Mho'.

The term 1/R is simply resistance divided into 1 giving the conductance in siemens. Taking any quantity 'x' and dividing that quantity into 1 is called the reciprocal of the quantity - this you will need to remember.

RESISTORS

Pictured are some of the types and sizes that resistors are packaged in. They are all just resistors. Resistors can be made to vary; these are called rheostats, potentiometers, or just plain variable resistors.

The resistors in figure 3-3 are not shown to scale.

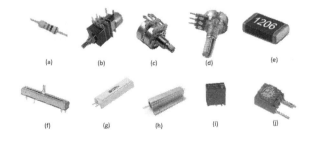

(a)   (b)   (c)   (d)   (e)

(f)   (g)   (h)   (i)   (j)

(a) fixed low wattage carbon film resistor

(b) a variable motorised resistor.

(c) two potentiometers on the one shaft.

(d) a variable resistor - potentiometer.

(e) a surface mount resistor

(f) a variable 'slider' resistor.

(g) high wattage wire wound resistors.

(h) a very high-power resistor with heat sink.

(i) a multi-turn circuit board mounted variable resistor.

(j) a circuit board mounted variable resistor - trim pot.

The only reason that resistors are made large is so they can dissipate (give off to their surroundings) heat. I only add this because a physically large resistor does not mean a large resistance. The large resistor in figure 3-3(h) may only be a few ohms and the tiny one in figure 3-3(a) could be 1MΩ. Large physical resistors will often not use the colour code and have their value written on them in ohms.

SCHEMATIC SYMBOLS OF RESISTORS

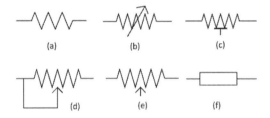

SCHEMATIC SYMBOLS OF RESISTORS

(a) fixed resistor.

(b) variable resistor.

(c) a trimmer potentiometer.

(d) rheostat - essentially the same as (e).

(e) a potentiometer.

(f) an alternative symbol for a fixed resistor.

4 - Ohm's Law

Ohm's Law describes the relationship between current, voltage and resistance in an electric circuit.

Ohm's Law states:

The current in a circuit is directly proportional to voltage and inversely proportional to the resistance.

Let:

I = current
E = voltage
R = resistance

Part of Ohm's Law says: current is directly proportional to voltage.

Using the symbols given, we can write an equation to show a direct proportion between current and voltage.

I=E

Normally the above equation is read I 'equals' E. It can just as easily and more understandably be read as: I is directly proportional to E.

I know I harp on about the direct proportion and inverse proportion stuff a lot. I do so because it is so important to understand this thoroughly before we come to do more complex equations.

I=E

Means that if the voltage is increased or decreased in a circuit, then the current will increase or decrease by the same amount. Double the voltage and you double the current. Halve the voltage and you halve the current. This is a direct proportion.

The other part of Ohm's Law says that current is inversely proportional to the resistance. This can be written as:

I=1/R

Now 1/R is a fraction with a numerator (the top part, 1) and a denominator, R. 1/R is a fraction just like 1/4, 1/2 and 3/8 are fractions.

R is the denominator in the fraction. What happens to the whole fraction if the denominator is changed? Watch.

1/2, 1/3, 1,4, 1/5, 1/6

As the denominator increases, the fraction decreases. In fact, if the denominator doubles then the fraction is half the size. 1/4 is half the size of 1/2.

'I' is the same as 1/R. This is an inverse proportion. If I is the same as 1/R and R is increased in size by three times, then the fraction 1/R is a third the size now and since 1/R is the same as the current, then the current is a third the size also.

The complete equation for Ohm's Law then is:

$$I = \frac{E}{R}$$

This equation, derived from Ohm's law, enables us to find the current flowing in any circuit if we know the voltage (E) and resistance (R) of the circuit.

For example, a resistor of 20 ohms has a 10 volt battery connected across it. How much current will flow through the resistor?

I=E/R =10/20 = 1/2 = 0.5 amperes

The equation I=E/R can be transposed for E or for I.

In some texts a diagram called the Ohm's Law triangle is used to help you rewrite the equation for E and R. This is okay, however, you do really need to know how to transpose equations - not just this one. If you learn to transpose this equation, then you will be able to do it with many others. There is a memory wheel at the end of this chapter that can help you remember equations for Ohm's law and power. You will also find a tutorial on transposing equations and using a calculator in the RES downloads area if you feel you might need some extra help. Always write to your facilitator if you require assistance.

We want to transpose I=E/R for E and R. The rule is: do whatever you like to the equation and it will always be correct as long as you do the same to each side of the equal sign. For example, if I multiply both sides of the equal sign by R, we get:

$$I \: x \: \frac{R}{1} = \frac{E}{R} \: x \: \frac{R}{1}$$

On the left-hand side (LHS) we have I x R. On the RHS we have E multiplied by Rand divided by R. Can you see that the R's cancel on the RHS? R/R is 1/1.

$$I \: x \: \frac{R}{1} = \frac{E}{1} \: x \: \frac{1}{1}$$

30

There is no need to show the 1s at all since multiplying or dividing a number by 1 does not change the number, therefore:

$$I \times R = E$$

Re-writing the above with E on the LHS we get:

$$E = IR$$

When there is no sign between two letters in an equation, like IR above, it is assumed the IR means I x R.

Now transpose the equation for R:

$$I = \frac{E}{R}$$

Multiply both sides by 1/E (which is the same as dividing both sides by E):

$$I \times \frac{1}{E} = \frac{E}{R} \times \frac{1}{E}$$

On the RHS the E's cancel out so we can rewrite the equation as:

$$I \times \frac{1}{E} = \frac{1}{R} \times \frac{1}{1}$$

or

$$\frac{I}{E} = \frac{1}{R}$$

Turning both sides upside down (remember we can do anything as long as we do the same to both sides):

$$\frac{E}{I} = \frac{R}{1}$$

Remove the '1' and reverse the sides to get:

$$R = \frac{E}{I}$$

So the three equations are:

$$I = \frac{E}{R} \qquad E = IR \qquad R = \frac{E}{I}$$

I have probably bored you by now - however; it is really important to be able to transpose equations for yourself. For one thing, you don't need to remember so many equations.

So, if you know any two of the three in 'E', 'R' and 'I' then you can transpose like we did here to calculate the missing one.

31

Finding I when you know E and R:

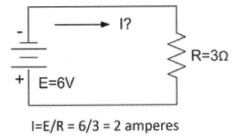

I=E/R = 6/3 = 2 amperes

Finding E when you know I and R:

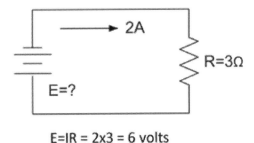

E=IR = 2x3 = 6 volts

Finding R when you know I and E:

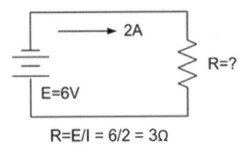

R=E/I = 6/2 = 3Ω

POWER

The unit of electrical power is the Watt (W), named after James Watt. One watt of power equals the work done in one second by one Volt of potential difference in moving one Coulomb of charge.

Remember that one coulomb per second is an ampere. Therefore, power in watts equals the product of amperes times volts.

Power (P) in watts= volts x amperes

P =E x I

Example: A toaster takes 5A from the 240V power line. How much power is used?

P = E x I = 240V x 5A
P = 1200 watts

Example: How much current flows in the filament of a household 75-watt light bulb connected to the standard 240 volt supply?

You know P (power) and E (volts). You need to transpose P=EI for 'I' and you get:

I=P/E Therefore:
I=75/240
I=0.3125 amperes

This amount of current is best expressed in milliAmperes. To convert amperes to milliamperes multiply by 1000 or think of it as moving the decimal point 3 places to the right, which is the same thing. This gives:

312.5mA Power in watts can also be calculated from:

$P=I^2R$, read, "power equals I squared R".
$P=E^2/R$, read, "power equals E squared divided by R".
Watts and Horsepower Units. 746W=1 horsepower.

This relationship can be remembered more easily as 1 horsepower equals approximately 3/4 kilowatt. One kilowatt= 1000W.

WORK

Work= Power x Time

Practical Units of Power and Work. Starting with the watt, we can develop several other important units. The fundamental principle to remember is that power is the time rate of doing work, while work is the power used during a period. The formulas are:

Power= work/time and
Work= power x time

The unit of power is the watt. One watt used during one second equals the work of one joule. In other words, one watt is one joule per second. Therefore, 1W = 1J/s. The joule is a basic practical unit of work or energy.

A unit of work that can be used with individual electrons is the Electron-Volt. Note: that the electron has charge, while the volt is potential difference. Now 1eV is the amount of work required to move an electron between two points having a potential difference of one volt. Since 6.25×10^{18} electrons equal 1C and a joule is a volt-coulomb, there must be 6.25×10^{18} eV in 1J.

Kilowatt-hours. This is a unit commonly used for large amounts of electrical work or energy. The amount is calculated simply as the product of the power in kilowatts multiplied by the time in hours during which the power is used. This is the unit of energy you need to know.

Example: A light bulb uses 100W or 0.1kW for 4 hours (h), the amount of energy used is:

Kilowatt-hours= kilowatts x hours

= 0.1 X 4
= 0.4kWh.

We pay for our household electricity in kilowatt-hours of energy.

POWER DISSIPATION IN RESISTANCE

When current flows through a resistance, heat is produced due to the energy consumed in moving free electrons and the atoms obstructing the path of electron flow. The heat is evidence that power is used in producing current. This is how a fuse opens, as heat resulting from excessive current melts the metal link in the fuse.

The power is generated by the source of applied voltage and consumed in the resistance in the form of heat. As much power as the resistance dissipates in heat must be supplied by the voltage source; otherwise, it cannot maintain the potential difference required to produce the current.

Any one of the three formulas can be used to calculate the power dissipated in a resistance. The one to be used is just a matter of convenience, depending on which factors are known.

In the following diagram, the power dissipated with 2A through the resistance and 6V across it is 2x6=12W. Or, calculating in terms of just the current and resistance, we get 2^2 times 3, which equals 12W. With voltage and resistance, the power can be calculated as 6x2 or 36 divided by 3, which also equals 12W.

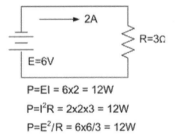

Resistance is the "only" electrical property that dissipates heat

We have introduced a new schematic symbol here too. The schematic symbol of a battery is shown at the left. Note the small bar at the top is the negative terminal. The direction of current (electron) flow is shown correctly, from negative to positive.

No matter which equation is used, 12W of power is dissipated, in the form of heat. The battery must generate this amount of power continuously in order to maintain the potential difference of 6V that produces the 2A current against the opposition of 3Ω.

In some applications, the electrical power dissipation is desirable because the component must produce heat in order to do its job. For instance, a 600W toaster must dissipate this amount of power to produce the necessary amount of heat. Similarly, a 300W light bulb must dissipate this power to make the filament white hot so that it will have the incandescent glow that furnishes the light. In other applications, however, the heat may be just an undesirable by-product of the need to provide current through the resistance in a circuit. In any case, though, whenever there is current in a resistance, it dissipates power equal to I^2R.

The term I^2R is used many times to describe unwanted resistive power losses in a circuit. You will hear of the expression I^2R losses as we go through this course.

ELECTRIC SHOCK

While you are working on electric circuits, there is often the possibility of receiving an electric shock by touching the "live" conductors when the power is on. The shock is a sudden involuntary contraction of the muscles, with a feeling of pain caused by current through the body. If severe enough, the shock can be fatal.

The greatest shock hazard is from high voltage circuits that can supply appreciable amounts of power. The resistance of the human body is also an important factor. If you hold a conducting wire in each hand, the resistance of the body across the conductors is about 10,000 to 50,000Ω. Holding the conductors tighter lowers the resistance. If you hold only one conductor, your resistance is much higher. It follows that the higher the body resistance, the smaller the current that can flow through you. Dry skin has much higher resistance than wet skin.

A safety rule, therefore, is to work with only one hand if the power is on. Also, keep yourself insulated from earth ground when working on power line circuits since one side of the line is usually connected to earth. Also, the metal chassis of radio and television receivers is often connected to the power line ground for your safety. The final and best safety rule is to work on the circuits with the power disconnected if at all possible and make resistance tests.

Note that it is current through the body, not through the circuit, that causes the electric shock. This is why care with high voltage circuits is more important since sufficient potential difference can produce a dangerous amount of current through the relatively high resistance of the body. For instance, 500V across a body resistance of 25,000Ω produces 0.02A, or 20mA, which can be fatal. As little as 10μA through the body can cause an electric shock. In an experiment on electric shock to determine the current at which a person could release the live conductor, this value of "let go" current was about 9 mA for men and 6 mA for women.

In addition to high voltage, the other important consideration in how dangerous the shock can be is the amount of power the source can supply. The current of 0.02A through 25,000Ω means the body resistance dissipates 10W. If the source cannot supply 10W, its output voltage drops with the excessive current load. Then the current is reduced to the amount corresponding to how much power the source can produce.

In summary, then, the greatest danger is from a source having an output of more than about 30V with enough power to maintain the load current through the body when it is connected across the applied voltage.

RESISTANCE OF EARTH

The earth, no not the ground, I am speaking of planet earth, is not made of metal (in any great concentrated amount) so one may expect that it is not a good conductor. However, if you recall the equation R = pl/ A, where A is the cross-sectional area - well the earth indeed does have an enormous cross-sectional area. This means for many applications the earth itself can be used as a conductor to save us having to run two conductors from the source to the load. Such circuits are called **earth return** and they have been used for power distribution and telephone communications.

Some Revision

By now you should have a good concept of current, voltage and resistances. It should be clear in your mind that current flows in a circuit pushed and/or pulled along by voltage. Current is restricted from flowing in a circuit by resistance.

You should be aware by now that statements like; "the voltage through the circuit" are in error. Voltage is electrical pressure. Voltage is never through anything. You can have voltage across the circuit or a component, but you can never have voltage through anything. Current flows through the circuit pushed along by voltage and restricted by resistance.

VOLTS PUSH AMPS THROUGH OHMS

A final point. You can have voltage without current. However, you cannot have current without voltage. A battery sitting on a bench has a voltage on its terminals - but no current is flowing. Voltage is electric pressure just like the water pressure in your tap. Current is the flow of electrons just like the flow of water from a tap. If the tap is turned off, you do not have a water flow. However, the pressure is definitely still there. Likewise, it is possible (with a disconnected battery) to have voltage (electric pressure) and no current (flow).

However, you cannot have any flow without pressure. So voltage can exist on its own, current cannot.

The unit of current is the Ampere. When 6.25×10^{18} electrons flow past a given point in a circuit in one second, the current is said to be one ampere.
Since 6.25×10^{18} electrons is a coulomb, this can be used in the definition of an ampere. An ampere of current is said to flow when one coulomb passes a given point in one second.

OHM'S LAW WHEEL

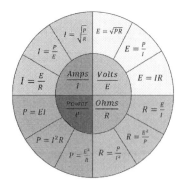

5 - Series Circuits

When components in a circuit are connected in serial order with the end of each joined up to the other end of the next as shown below in figure 5-1, they form a series circuit.

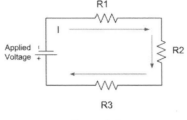

Figure 5-1

An electric current consists of an ordered movement of electrons. In the schematic shown in figure 5-1, the current leaves the negative terminal of the battery and flows through R1, R2 and R3, then returns to the positive terminal.

It does not matter where we measure the current in a series circuit as we will always get the same value of current everywhere, as:

Current is the same in all parts of a series circuit

The total resistance of any number of resistances in series is simply the sum of the individual resistances:

Rt=R1+R2+R3..n

Suppose the resistors in figure 5-1 were 10, 20 and 30 ohms respectively and the applied voltage was 10 volts. What is the current flowing in the circuit?

Rt=R1+R2+R3
Rt=10+20+30
Rt=60 ohms

We now know the total resistance (60 ohms) and the applied voltage is 10 volts, so we can use Ohm's law to calculate the current flowing in the circuit.

I=E/R = 10/60 = 1/6 A or 0.1666 A or 166.6 milliamps

All the resistances shown in the four circuits of figure 5-2 are in series. The circuits are drawn differently, but nevertheless, they are all series circuits and the current is the same in every part of the circuit. The rectangle symbol is an alternative method of drawing a resistor. In this circuit the rectangle could be any type of component.

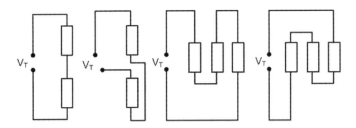

Another rule that you must learn is:

The sum of the voltage drops in a series circuit is equal to the applied voltage.

Going back to figure 5-1, we have three resistances R1, R2 and R3 in series, connected to a 10 volt supply. We can calculate the voltage across R1 because we know the resistance and we know the current through R1.

Let's call the voltage across R1 'ER_1',
then:

ER1 = I x R1 = 0.1666 x 10 = 1.666 volts
And for the other two resistances:
ER2 =I x R2 = 0.1666 x 20 = 3.332 volts
ER3 =I x R3 = 0.1666 x 30 = 4.998 volts

The sum of the voltage drops across each resistance will equal the applied voltage. There will be a small error in our example due to rounding of decimals.

$E_t = E_{R1} + E_{R2} + E_{R3}$ = 1.666 + 3.332 + 4.998 = 9.996 volts (with rounding error).

Notice how the supply voltage is distributed around the circuit - think about it for a while. The least resistance has the smallest voltage drop across it and the largest resistance has the most voltage across it. This makes sense when we realise that the current in a series circuit must be the same through all components. R3 needs nearly half of the supply voltage to get 0.1666 amps to flow through it and R1 requires only 1.666 volts to produce the same current through it.

If two resistances of equal value are connected in series across a supply voltage of say 20 volts, then each resistance would have exactly half the supply voltage across it. You can work out the voltage drops in a series circuit by using a method known as 'proportion'.

Let's use our series circuit again and this time, we will work out the voltage drops for R1, R2 and R3 without using the current.

The total resistance of the circuit is 60 ohms (R_t) and the supply voltage is 10 volts.

E_{R1} = R1/R_t x 10 = 10/60 x 10 = 1.6667 volts (rounded)
E_{R2} = R2/R_t x 10 = 20/60 x 10 = 3.3334 volts (rounded)
E_{R3} = R3/R_t x 10 = 30/60 x 10 = 5 volts

AN EXAMPLE OF A SERIES CIRCUIT

The best example I can think of for a common series circuit which will also demonstrate one of the problems with series circuits is Christmas tree lights. These lights are low voltage lights of about 10 volts. Christmas tree lights plug into the mains voltage of 240 volts. Suppose each bulb requires 10 volts and you connect 24 of them in series, then each bulb will have 10 volts across it.

At least this is the way Christmas tree lights are supposed to work. Darn dangerous things if you ask me because if you break a bulb you are exposing yourself to potentially (no pun intended) 240 volts.

So, from the point of view of Christmas tree lights, series circuits have an advantage. The disadvantage is, should a bulb blow, no current will flow in the circuit and all the lights will go out.

What would happen if you placed a short circuit where a bulb had blown (don't do this by the way)? The remaining 23 lights would come on again as you have completed the circuit. However, they would all be a little brighter than intended as they will now have slightly more than their designated 10 volts each. Of course running them at a higher voltage will cause them to burn out faster, so if you continued down this track and survived, you would find that the bulbs would blow faster and faster until eventually they would blow immediately.

POWER IN A SERIES CIRCUIT

The total power in a series circuit is found from P=EI where E is the applied voltage and I is the total current.

In our example: 10 x 0.1666 = 1.666 watts

You can use any of the power equations to calculate the power dissipated in R1, R2 or R3. Of course, the sum of the power dissipated in each resistance should equal the total power in the circuit (1.666 watts).

$P_{R1} = 1^2$ x R1 = $(0.1666)^2$ x 10 = 0.2775556 watts
$P_{R2} = 1^2$ x R2 = $(0.1666)^2$ x 20 = 0.5551112 watts
$P_{R3} = 1^2$ x R3 = $(0.1666)^2$ x 30 = 0.8326668 watts

$Pt= P_{R1} + P_{R2} + P_{R3}$
$Pt= P_{R1} + P_{R2} + P_{R3}$ = 1.6653336 watts (slight rounding error)

THE VOLTAGE DIVIDER

A voltage divider is a simple way to adjust the voltage in a particular part of a larger circuit. The supply voltage might be 9V and the component requires 2V on one terminal. A voltage divider can be used to produce the required 2 volts. Two resistors can be connected to produce such a voltage divider. Such a circuit is shown in figure 5-3.

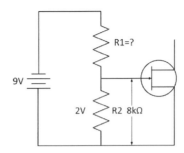

Since we have not looked at parallel circuits yet, we will assume that the total resistance of the transistor is 8000Ω and this is represented by the resistance R2. This is a real problem we will learn more about this later when we look at transistors. For now, R2 is 8000Ω and we want to workout the value of R1 that will produce 2 volts across R2.

Our problem then is to find the value of R1.

We know that the sum of the voltage drops across R1 and R2 must equal the supply voltage of 9 volts. R2 has 2 volts across it so R1 must have 7 volts across it. It should be evident that the resistance value of R1 must be higher than that of R2 since it has the higher voltage across it.

Let's try this: -

E_{R2} we know is 2 volts and R2 is 8000Ω.
We can calculate the current through R2:
$I_{R2} = E_{R2}/R2 = 2/8000 = 0.25$ milliamperes

We now know the current through R1 since R2 and R1 are in series.

$R2 = E_{R2}/I_{R2}$

In case the abbreviations are confusing you - E_{R2} means the voltage across R2 and I_{R2} means the current through R2.
R1= 7 volts/ 0.25 milliamperes ← remember this is milliamperes
R1 = 28000Ω

We can check this by saying:

8000/36000th of the supply voltage should be across R1,
or 8/36 x 9 which gives 2 volts which is correct.

Also:
28000/36000th of the supply voltage should be across R2,
We know that the sum of the voltage drops across R1 and R2 must equal the supply 28/36 x 9 = 7 volts.

In electronic circuits we often use voltage dividers to bias transistors or establish the correct operating voltage of a component when the supply voltage is too high.

6 - Parallel Circuits

When two or more components are connected directly across one voltage source, they form a parallel circuit. The two lamps in figure 6-1 are in parallel with each other and with the battery. Each parallel path is called a branch, with its own individual current. Parallel circuits have one common voltage across all the branches, but the individual branch currents can be different.

The voltage is the same across all components in a parallel circuit

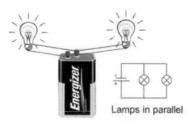

Lamps in parallel

Figure 6-1

In figure 6-1 a pictorial diagram and the schematic equivalent circuit is shown. The two lamps are directly connected to the battery terminals. This is always the case with parallel circuits. If you had 10 components (they don't have to be lamps) connected in parallel, then each side of each component is connected directly to the battery.

BRANCH CURRENTS

Each resistance (or other component) in a parallel circuit is connected by a conductor directly to the source voltage. Each resistor will draw current from the source according to Ohm's law, $I=E/R$, for each branch. The sum of all the branch currents must then be equal to the total current drawn from the source. In a parallel circuit the sum of the branch currents equals the total current.

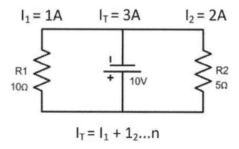

$$I_1 = 1A \qquad I_T = 3A \qquad I_2 = 2A$$

R1 10Ω   10V   R2 5Ω

$$I_T = I_1 + I_2...n$$

Figure 6-2

In applying Ohm's law, it is important to note that the total current equals the voltage applied across the circuit divided by the total resistance of the circuit. $I_T=E_T/R_T$.

In figure 6-2, 10V is applied across the 10Ω of R_1, resulting in a current of 1 Ampere being drawn from the battery through R_1. Similarly, the 10 volts applied to the 5Ω of R_2 will cause 2 amperes to be drawn from the battery.

The two branch currents in the circuit are then 1 ampere and 2 amperes. The total current drawn from the battery is then 3 amperes.

Just as in a circuit with only one resistance, any branch that has less resistance will draw more current. If R_1 and R_2 were equal, however, the two branch currents would have the same value. For instance, if R_1 and R_2 were both 50Ω then each branch would draw 2 amperes and the total current drawn from the battery would be 4 amperes.

The current can be different in parallel circuits having different resistances because the voltage is the same across all the branches. Any voltage source generates a potential difference across its two terminals. This voltage does not move. Only current flows around the circuit.

The source voltage is available to make electrons move around any closed path connected to the generator terminals. How much current flowing in the separate paths depends on the amount of resistance in each branch.

For a parallel circuit with any number of branch currents we can then write an equation for calculating the total current (I_T):

$I_T=I_1+I_2+I_3..n$

This rule applies to any number of parallel branches, whether the resistances are equal or unequal.

Example:

An R_1 of 20Ω and an R_2 of 40Ω and an R_3 of 60Ω are connected in parallel across a 240 volt supply. What is the total current drawn from the supply?
Let's calculate the branch currents for R_1, R_2 and R_3:
$IR_1 = E/R_1 = 240/20 = 12$ amps
$IR_2 = E/R_2 = 240/40 = 6$ amps
$IR_3 = E/R_3 = 240/60 = 4$ amps

The total current drawn from the 240 volt supply is the sum of the branch currents:

$I_T= IR_1 + IR_2 + IR_3 = 12 + 6 + 4 = 22$ amperes.

RESISTANCES IN PARALLEL

In the example, above we could have worked out the total resistance to calculate the total current being drawn from the supply.

To find the total resistance of any number of resistors in parallel we find the reciprocal of the sum of the reciprocals for each resistance. This sounds like a bit of a mouthful so I will put it in equation form and you should see what I mean.

$$R_t = \frac{1}{1/R_1 + 1/R_2 + 1/R_3}$$

Let's calculate the total resistance of our example using this equation.
Firstly, find the reciprocal of each of the resistances:

Reciprocal of R_1 = $1/R_1$ = $1/20$ = 0.05
Reciprocal of R_2 = $1/R_2$ = $1/40$ = 0.025
Reciprocal of R_3 = $1/R_3$ = $1/60$ = 0.01667 (recurring decimal)

The sum of the reciprocals above is: 0.091667

Finally, to find Rt we take the reciprocal of the sum of the reciprocals, or:

1/0.091667 which equals 10.91 rounded.

So Rt= 10.91 ohms (with a little rounding error).
Since we know the total resistance and the applied voltage we can now calculate the total current from Ohm's law.

I=E/R = 240/10.91 = 21.998 amps, or close enough to the 22 amps we calculated earlier. When you get used to it, these calculations are very easy to do on a calculator. Many calculators have a reciprocal key '1/x' or 'x^{-1}'.

Example: A parallel circuit consisting of two branches, with a current of 5A through each branch, is connected across a 90V source. What is the equivalent total resistance Rt?

To find the total resistance (Rt) we need to know the applied voltage (90V) and the total current drawn from the supply. Since we are told that there are two branches and each branch draws 5A then the total current must be 10A. We can now use Ohm's Law to calculate the total resistance:

R=E/I
R=90/10
R=90Ω

PARALLEL BANK

A combination of parallel branches is often called a bank. Suppose you needed a resistance of 8Ω for the load of a high power, 75 Watt, audio amplifier. It is likely that you do not have an 8Ω 75W resistor. However, let's say you have a draw full of 120Ω 5W resistors. You could connect 15 x 120Ω 5W resistors in parallel to give a resistive load of 8Ω 75W.

120/15=8Ω, and the power is 5W x 15 = 75W. If you are going to use the parallel resistive bank as a 75W load for a long time you will need to think about adding cooling. This could be a fan or you could even immerse the resistor bank in a metal can filled with non-corrosive oil.

A stumbling block for many is trying to understand how adding more resistors to a parallel circuit can reduce the total resistance.

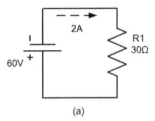

Figure 6-3(a)

In figure 6-3(a) you see a supply voltage of 60 volts connected to a 30Ω load. By Ohm's law the current drawn by the 30Ω load must be 2A.

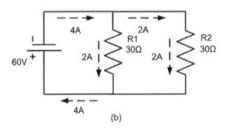

Figure 6-3(b)

If a second 30Ω resistor is now connected as in figure 6-3(b), then an additional 2A will flow through it. The total current drawn from the supply is now 4A.

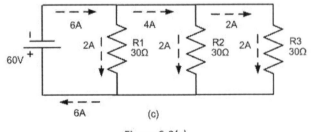

Figure 6-3(c)

Adding a third 30Ω resistor as in figure 3(c) causes a further 2A to be drawn from the supply, bringing the total current drawn from the supply to 6A.

The total resistance of this circuit would be:

$R = E/I = 60/6 = 10\,\Omega$

We can see that as we add parallel resistors to the supply, more current is drawn from the supply, so the total circuit resistance must be less. We could go on adding parallel resistors for as long as we wanted and each time we did, the total circuit resistance would become less with each added resistor.

So now you should be able to answer the question; why does adding more resistors to a parallel circuit increase the total current? Each additional parallel resistor creates an additional branch for current that would not otherwise be there.

DERIVING THE RECIPROCAL EQUATION

We discussed the reciprocal equation earlier, but we did not mention how this equation was derived. We know the basic law: the sum of the branch currents is equal to the total current flowing in a parallel circuit.

$I_t = E/R_t$ and $I_1 = E/R_1$ and $1_2 = E/R_2 \dots n$

We can substitute this into the rule for branch currents and get:

$E/R_t = E/R_1 + E/R_2 + E/R_3 \dots n$

Notice how E is in the numerator on both sides of the equal sign. If we divide both sides by E, the E's cancel out and we are left with:

$1/R_t = 1/R_1 + 1/R_2 + 1/R_3 \dots n.$

This gives us the reciprocal of the total resistance. The reciprocal of this is the total resistance.

SPECIAL CASE OF ALL RESISTANCES THE SAME

If all of the resistances are the same in a parallel circuit, as in our dummy load example presented earlier, the total resistance can be found by dividing the number of branches into the value of one of the parallel resistances. In the dummy load example, we had twenty, $1000\,\Omega$ resistors. A quick method to find the total resistance is; $1000/20 = 50\,\Omega$. However, please do remember that this approach can only be used when all resistances are the same.

When there are two parallel resistances and they are not equal, it is usually quicker to calculate the combined resistance by the method known as 'product over sum'. This rule says that the combination of two parallel resistances is their product divided by their sum.

$R_t = (R_1 \times R_2)/(R_1 + R_2)$

The symbol '/' means divided by.

Example:

A resistance of 100Ω is connected in parallel with a resistance or 200Ω. What is the total resistance of the circuit using the product over sum method?

Rt = (100x200)/(100+200)
Rt = 20000 / 300
Rt= 66.6670Ω (rounded).

Notice also that the total resistance of parallel resistors is always less than the lowest resistance.

TOTAL POWER IN PARALLEL CIRCUITS

Since the power dissipated in the branch resistances must come from the source voltage, the total power equals the sum of the individual values of power in each branch.

Pt=PR1+PR2+PR3..n

Of course if you know the total current and the applied voltage, you can find the power using P=EI.

I can recall a question that was asked in a electrical exam some years ago. I am sure the question was not meant to be as difficult as it turned out and my guess is that the person writing the question must have been anxious to knock off work at the time.

The question was:

Four 10Ω, 1-watt resistors are connected in series. What is the total power dissipation rating of the circuit?

That was it, no voltage or current mentioned. In fact, the resistance (10Ω) is totally irrelevant as is the series connection.

The answer is 4 watts.

The question was poorly phrased in my view and could have been put:
What is the maximum power that four, 1 Watt resistors can dissipate?

ADVANTAGE OF A PARALLEL CONNECTION

Household appliances, as you may know, are designed to operate at 240V in Australia and many other countries. The electrical wiring to all the power points and light fittings in your home are in the form of a parallel circuit. This ensures that each appliance receives the supply voltage of 240V. Imagine the problems if your house was wired as a series circuit. Each time you turned on a light all the lights in the house would get dimmer. If you turned on something with a high resistance (low wattage), the lights would not even glow. Supplying the same voltage to various devices in a house, a radio circuit, or a car requires parallel connections.

7 - Series Parallel Circuits

In many circuits, some components are connected in series, so they have the same current, while others are in parallel for the same voltage. When analysing and doing calculations with series-parallel circuits, you simply apply what you have learned from the last two chapters.

In the circuit of figure 7-1, we could work out all the voltages across all of the resistances and the current through each resistance and then total resistance. For now, I am just going to walk through the simplification of this circuit to a single resistance connected across the 100V source.

Keep in mind that any circuit (resistive) can be reduced to a single resistance. This is particularly useful when we come to do transmission lines and antennas.

For now, let's have a go at simplifying the circuit of figure 7-1. There are many ways to go about this problem. The method I prefer is to start at the right-hand side and work my way back to the source, simplifying the circuit as I go. Let's try this together: -

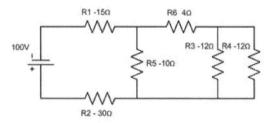

Figure 7-1

On the right-hand side, we see R3 and R4 in parallel and each 12Ω. Do you remember the shortcut method when parallel resistances are all the same value? Divide the value of the branch by the number of branches: 12Ω /2 = 6Ω.

Replace the parallel pair of R3 and R4 with a single resistance R7 as shown in figure 7-2.

Always redraw the circuit.

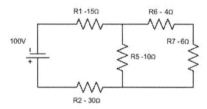

Figure 7-2

The next logical step would be to combine the series resistance of R6 and R7 into a single resistance R8 by adding them, as shown in figure 7-3: 4Ω + 6Ω = 10Ω.

Redraw the circuit.

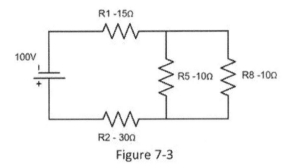

Figure 7-3

Now combine the parallel and equal resistances of R5 and R8 into a single resistance R9 of 5Ω as shown in figure 7-4.

Redraw the circuit.

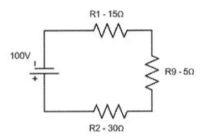

Figure 7-4

We are now left with a series circuit consisting of R1, R2 and R9. We find the total resistance of these by adding them: 15Ω + 30Ω + 5Ω =50Ω.

You guessed it! **Redraw the circuit** again for our final result shown in figure 7-5.

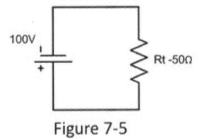

Figure 7-5

So, the total resistance of the circuit we started with is equivalent to a single resistance of 50Ω.

We could now find out how much current is being drawn by the circuit from the supply, by using Ohm's law:

I=E/R = 100/50 = 2 amperes.

That's it! You will find that most circuits, even very complex ones, can be handled in the same manner i.e. by simplifying series and parallel branches as you work your way down to a single resistance.

I cannot emphasise enough the need to redraw the circuit as you work your way through it. The calculations for this circuit were easy arithmetically. If you found it was not so easy, then it is important for you to show all of your working out as well.

Refer to the circuit of figure 7-1. What is the voltage across R4?

We know that the total resistance of the circuit is 50Ω and from this we worked out that the current drawn from the supply was 2A.

Therefore, if we go back to the figure 7-1 circuit, R1 (15Ω) and R2 (30Ω) must have 2A flowing through them since they are in series with the supply.

The voltage across R1 will be: $E=IR_1 = 2 \times 15 = 30V$.

Likewise, the voltage across R2 will be: $E=IR_2 = 2 \times 30 = 60V$.

Now if there are 30 volts across R1 and 60 volts across R2 then this leaves 100-30-60 = 10 volts across R5.

R5 is in parallel with R6 and the parallel pair of R3 and R4.

R3 and R4 simplify to 6Ω.

So we have 10 volts across 4Ω and 6Ω in series. The voltage across the 6Ω will be the voltage across R4 (and R3 for that matter).

Some may see immediately without any calculation that 10 volts across a series combination of a 4Ω and a 6Ω resistor will result in 4 volts across the 4Ω and 6 volts across the 6Ω. If you can't see this then don't worry, let's solve it using Ohm's law.

$I = E/R = 10/10 = 1$ ampere.

So, there is 1 ampere flowing through the 6Ω resistance that represents the combined resistance of R3 and R4 in parallel.

$E = IR = 1 \times 6 = 6$ volts.

Therefore, the voltage drop across R4 is 6 volts.

SERIES CIRCUIT	PARALLEL CIRCUIT
The current in all parts of the circuit is the same.	The voltage is the same across all parallel branches.
E across each series R is I x R.	I in each branch is E/R.
The sum of the voltage drops equals the applied voltage. $E_t = E_1 + E_2 + E_3$ etc.	The sum of the branch currents equals the total current. $I_t = I_1 + I_2 + I_3$ etc.
$R_t = R_1 + R_2 + R_3$ etc.	$1/R_t = 1/R_1 + 1/R_2 + 1/R_3$ etc.
R_t must be larger than any individual R.	R_t must be less than the smallest branch R.
$P_t = P_1 + P_2 + P_3$ etc.	$P_t = P_1 + P_2 + P_3$ etc.
Applied voltage is divided into IR drops.	Main current is divided into branch currents.
The largest voltage drop is across the largest resistance.	The largest branch current is through the smallest parallel R.
Open circuit in one component causes the entire circuit to be open.	Open circuit in one branch does not prevent current in other branches.

8 - Magnetism

MAGNETISM AND ELECTRICITY

Any wire carrying a current of electrons is surrounded by area of force called a magnetic field. For this reason, any study of electricity or electronics must consider magnetism.

Almost everyone has had experiences with magnets or pocket compasses at one time or another. A magnet attracts pieces of iron but has little effect on practically everything else. Why does it single out iron? A compass, when laid on a table, swings back and forth, finally coming to rest pointing toward the North Pole of the world.

Why does it always point in the same direction?

These and other questions about magnetism have puzzled scientists for hundreds of years. It is only relatively recently that theories have begun to answer many of the perplexing questions about magnetism.

Electrical and electronic apparatus such as relays, circuit breakers, earphones, loudspeakers, transformers, chokes, magnetron tubes, television tubes, phonograph pickups, tape and disk recorders, microphones, meters, motors and generators depend on magnetic effects to make them function.

Every coil (inductance) in an electric circuit is utilising the magnetic field that surrounds it when current is flowing through it. But what is meant by the term magnetic field?

THE MAGNETIC FIELD

An electron at rest has a negative electrostatic field of force surrounding it. When energy is imparted to an electron to make it move, a new type of field develops around it, at right angles to its electrostatic field. Whereas negative electrostatic lines of force are considered as radiating outward from an electron, the electromagnetic field of force develops as a ring around a moving electron, at right angles to the path taken, around the wire.

An interesting difference between a magnetic field and an electrostatic field is that an electrostatic field can exist around a single stationary charge such as an electron or a proton. This is not the case with magnetism. A North 'monopole' does not exist any more than a South 'monopole' exists. With magnetism, you always have two poles.

Electrons orbiting around the nucleus of an atom or a molecule produce electromagnetic fields around their paths of motion. These fields (electric and magnetic) are balanced or neutralised by the effect of proton movement in the nucleus. The movement of one orbital electron is counteracted by another orbital electron whirling in an opposite direction. In almost all substances, the net result is little or no external fields.

In the case of an electric conductor carrying current, the collective movement of electrons along the wire produces a magnetic field around the conductor. The greater the current, the more intense the magnetic field.

The diagram in figure 8-1 shows a method of determining the direction of the magnetic field around a current carrying conductor.

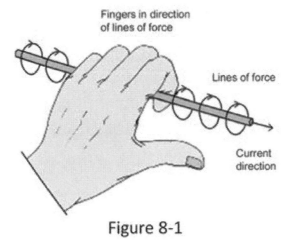

Figure 8-1

If you place a compass next to a conductor and then pass a current through the conductor, the compass needle will move indicating the presence of a magnetic field.

Under normal circumstances, the field strength around a current carrying conductor varies inversely as the distance from the conductor. At twice the distance from the conductor, the magnetic field strength is one-half as much, at five times the distance the field strength is one-fifth and so on. At a relatively short distance from the conductor, the field strength may be quite weak.

When the current in a conductor is increased, more electrons flow, the magnetic field strength increases and the whole field extends further outward.

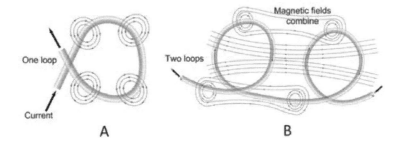

Figure 8-2

By looping a conductor, as shown in figure 8-2A, magnetic lines of force are concentrated in the central core area of the loop.

When two or more turns of wire are formed into a coil, as shown in figure 8-2B, the lines of force from each turn link to the fields of the other turns and a more concentrated magnetic field is produced in the core of the coil.

By forming a coil into many loops the magnetic field is stronger and more concentrated. With enough turns plus current a strong electromagnet can be created.

An experiment to try

Typically, in face-to-face classes, I would demonstrate many things by experiment. This makes learning more enjoyable. We don't have that luxury. However, I will mention these experiments from time to time and perhaps you can try them yourself. If you have never made an electromagnet, I really encourage you to do so as it can be a lot of fun and fascinating.

Get an iron rod about 75mm long. A bolt with a head and a nut screwed onto one end is ideal. Wind as many turns of wire onto it as you can. I use single telephone wire. However, any small diameter wire will do the job. The more turns, the better. Connect the ends of the coil you have made to a 6 volt lantern battery and you will have yourself an electromagnet. You can use the electromagnet with other experiments down the track.

Besides just playing with your electromagnet you can use it to magnetise some tools such as screwdrivers etc. Don't leave it connected to the battery for very long as it draws a heavy current from it. Do not use a car battery as the coil will overheat and melt. The coil (inductor) is a very low resistance. However, current is limited by using a 6 volt lantern battery, which can only supply 3-4 amps. If you want to use a car battery you need to limit the current flow through the electromagnet. You could do this by connecting a 12 volt automotive light bulb in series with the magnet.

Figure 8-3 shows an electromagnet made from a bolt and some scrap wire; a permanent bar and horseshoe magnet.

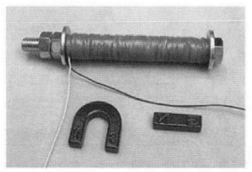

Figure 8-3

The direction of the field of force (north and south pole) can be reversed by reversing the current direction or by reversing the winding direction.

At one end of the coil the field lines are leaving and at the other end, they are entering. When a coil or piece of metal has lines of force leaving one end of it, that end is said to have a north pole. The end with the lines entering is the south pole.

The terms "north" and "south" indicate magnetic polarity, just as "negative" and "positive" indicate electrostatic polarity. They should not be used interchangeably.

The negative end of a coil is the end connected to the negative terminal of the source and does not refer to the north or south magnetic polarity of the coil.

All magnetic lines of force are complete loops and may be considered somewhat similar in their action to stretched rubber bands. They will contract back into the circuit from which they came as soon as the force that produced them ceases to exist.

Magnetic lines of force never cross each other. When two lines of force have the same direction, they will oppose mechanically if brought near each other.

PERMEABILITY

When a coil of wire is wound with air as the core, a certain flux density will be developed in the core for a given value of current. If an iron core is inserted into the coil, the flux density will increase. This increase in flux density is achieved without increasing current or the number of turns.

With an air core coil, the air surrounding the turns of the coil may be thought of as pushing against the lines of force and tending to hold them close to the turns. With an iron core, the lines of force find a medium in which they can exist much more easily than in air. As a result, lines that were held close to the turns in the air core coil are free to expand into the highly receptive area afforded by the iron. This allows lines of force that would have been close to the surface of the wire to expand into the iron core. Thus, the iron core produces a greater flux density, although there are no extra magnetic lines of force. In other words, for the same number of turns and current, a coil will have a stronger magnetic field if an iron core is inserted. The iron core merely brings the lines of force out where they can be more readily used and concentrates them.

The ability of material to concentrate magnetic lines of force is called Permeability.

The permeability of most substances is very close to that of air, which may be considered as having a value of 1. A few materials, such as iron, nickel and cobalt, are highly permeable, with permeabilities of several hundred to several thousand times that of air. The word "permeability" is a derivation of the word "permeate," meaning "to pervade or saturate" and is not related to the word "permanent.")

Permeability is represented by the Greek letter μ (mu, pronounced mew).

Alloying iron makes it possible to produce a broad range of permeabilities. Most stainless steel exhibits practically no magnetic effect, although some steels may be magnetic.

Any substance that is not affected by magnetic lines of force and is reluctant to support a magnetic field is said to have the property of reluctance. Air, vacuum and most substances have unity reluctance, while iron has a very low reluctance.

In electric circuits, the reciprocal of resistance is called conductance. In magnetic circuits, the reciprocal of reluctance is called permeance. "Permeability" is used when discussing how magnetic materials behave.

THE ATOMIC THEORY OF MAGNETISM

The discussion here will be a considerably condensed version of the atomic theory of magnetism.

From atomic theory, it is known that an atom is made up of a nucleus of protons surrounded by one or more electrons encircling it. The rotation of electrons and protons in most atoms is such that the magnetic forces cancel each other. Atoms or molecules of the elements iron, nickel and cobalt, arrange themselves into magnetic entities called domains. Each domain is a complete miniature magnet.

Groups of domains form crystals of the magnetic material. The crystals may or may not be magnetic, depending on the arrangement of the domains in them. Investigation shows that while any single domain is fully magnetised, the external resultant of all the domains in a crystal may be a neutral field.

Randomised domains   Aligned domains

Figure 8-4

Each domain has three directions of magnetisation: easy magnetisation, semi-hard magnetisation and hard magnetisation. If an iron crystal is placed in a weak field of force, the domains begin to line up in the easy direction. As the magnetising force is increased, the domains begin to roll over and start to align themselves in the semihard direction. Finally, as the magnetising force is increased still more, the domains are lined up in the hard direction. When all the domains have been lined up in the hard direction, the iron is said to be saturated. An increase in magnetising force no longer magnetically changes the material. The material is magnetically saturated.

FERROMAGNETISM

Substances that can be made to form domains are said to be ferromagnetic, which means "iron magnetic." The ferromagnetic elements are iron, nickel and cobalt, but it is possible to combine some non-magnetic elements and form a ferromagnetic substance. For example, in the proper proportions, copper, manganese and aluminium, each by itself being non-magnetic, produce an alloy which is similar to iron magnetically.

Materials made up of non-ferromagnetic atoms, when placed in a magnetic field may weakly attempt either to line up in the field or to turn at right angles to it. If they line themselves in the direction of the magnetic field, they are said to be paramagnetic. If they try to turn from the direction of the field, they are called diamagnetic. There are only a few diamagnetic materials. Some of the more common are gold, silver, copper, zinc and mercury. All materials which do not fall in the ferromagnetic or diamagnetic categories are paramagnetic. The greatest percentage of substances are paramagnetic.

Ferromagnetic substances will resist being magnetised by an external magnetic field to a certain extent. It takes some energy to rearrange even the easy to move domains.

Once magnetised, however, ferromagnetic substances may also tend to oppose being demagnetised. They are said to have retentivity, or remanence, the ability to retain magnetism when an external field is removed.

As soon as the magnetising force is released from a magnetised ferromagnetic substance, it tends to return at least part way back to its original non-magnetised state, but it will always retain some magnetism. This remaining magnetism is residual magnetism. Paramagnetic and diamagnetic materials always become completely nonmagnetic when the external magnetising force is removed from them.

PERMANENT AND TEMPORARY MAGNETS

Ferromagnetic substances that hold magnetic domain alignment well (have a high value of retentivity) are used to make permanent magnets. One of the strongest permanent magnets is made of a combination of iron, aluminium, nickel and cobalt called Alnico. It is used in horseshoe magnets, electric meters, headphones, loudspeakers, radar transmitting tubes and many other applications. Some magnetically hard, or permanent magnetic materials, are cobalt steel, nickel aluminium, steels and special steels.

Figure 8-5 shows the magnetic field surrounding permanent magnets. This picture was made by placing a piece of paper over the magnets and sprinkling iron filings onto the paper.

You can see the lines of force attracting when a north and a south pole are brought near to each other and repelling when two north poles are brought near to each other.

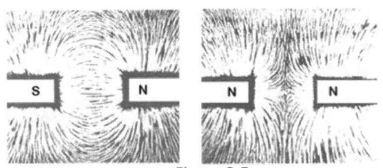

Figure 8-5

Ferromagnetic metals that lose magnetism easily (have a low value of retentivity) make temporary magnets. They find use in transformers, chokes, relays and circuit breakers.

Pure iron and permalloys ('perm' derived from "permeable," not from "permanent") are examples of magnetically soft, temporary magnet materials. Finely powdered iron, held together with a non-conductive binder, is used for cores in many applications. These are called ferrite cores.

MAGNETISING AND DEMAGNETISING

There are two simple methods of magnetising a ferromagnetic material. One is to wrap a coil of wire around the material and force a direct current through the coil. If the ferromagnetic material has a high value of retentivity, it will become a permanent magnet.

If the material being magnetised is heated and allowed to cool while subjected to the magnetising force, a greater number of domains will be swung into alignment, and a greater permanent flux density may result. Hammering or jarring the material while under the magnetising force also tends to increase the number of domains that will be affected.

A less efficient method of magnetising is to stroke a high retentivity material with a permanent magnet. Have you ever magnetised a screwdriver using this approach? This will align some of the domains of the material and induce a relatively weak permanent magnetism.

If a permanent magnet is hammered, many of its domains will be jarred out of alignment and the flux density will be lessened. If heated, it will lose its magnetism because of an increase in molecular movement that upsets the domain structure. Strong opposing magnetic fields brought near a permanent magnet may also decrease its magnetism. It is important that equipment containing permanent magnets be treated with care. The magnets must be protected from physical shocks, excessive temperatures and strong alternating or other magnetic fields.

When heated, permanent magnets lose their magnetism quickly and also at a certain temperature. The temperature at which a magnet loses its magnetism is called the Curie temperature.

When tools or objects such as screwdrivers or watches become permanently magnetised, it is possible to demagnetise them by slowly moving them into and out of the core area of a many turn coil in which a relatively strong alternating current (AC) is flowing. The AC produces a continually alternating magnetising force. As the object is placed into the core area, it is alternately magnetised in one direction and then the other. As it is pulled further away, the alternating magnetising forces become weaker. When it is finally out of the field completely, the residual magnetism will usually be so low as to be of no consequence.

PERMANENT MAGNET FIELDS

When a piece of magnetically hard material is subjected to a strong magnetising force the domains are aligned in the same direction. When the magnetising force is removed, many of the domains remain in the aligned position and a permanent magnet results. A north pole is anywhere the direction of the magnetic lines of force move outward from the magnet. A south pole is any place where the direction of the lines of force are inward.

If a magnet is completely encased in a magnetically soft iron box, all its lines of force remain on the walls of the box and there is no external field. This is known as magnetic shielding. Shielding may be used in the opposite manner. An object completely surrounded by an iron shield will have no external magnetic fields affecting it, as all such lines of force will remain in the permeable shield.

THE MAGNETISM OF THE EARTH

Sufficient quantity of the ferromagnetic materials making up the earth have domains aligned in such a way that the earth appears to be an enormous permanent magnet. The direction taken by the lines of force surrounding the surface of the earth is inward at a point near what is commonly known to be the north pole of the world and outward near the earth's south pole.

The simple magnetic navigational compass consists of a small permanent magnet balanced on a pivot point. The magnetic field of the compass needle lines itself up in the earth's lines of force.

Figure 8-6

As a result, the magnetic north end of the compass needle is pulled toward the earth's south magnetic pole, since unlike poles attract each other. This means that when the "north-pointing end" is pointing toward geographical north, this end (a magnetic north pole) is pointing toward a magnetic south pole.

RELAYS

A relay is a relatively simple magnetic device that generally consists of a coil, a ferromagnetic core and a movable armature on which make and/or break contacts are fastened. A simple relay may be used to close a circuit when the coil is energised. This type of relay is known as single pole single throw (SPST), normally open (NO), or "make contact" relay. The relay in figure 8-7 is a double pole single throw. The contacts can be used to turn something ON or OFF.

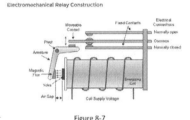

Figure 8-7

The core, of the relay and the pivoted armature bar, are made of magnetically soft ferromagnetic materials having high permeability and little retentivity. One of the relay contacts is attached by an insulating strip to the armature and the other to the relay body with an insulating material. The contacts are electrically separated from the operational parts of this particular relay.

When current flows through the coil, the core is magnetised and lines of force develop in the core and through the armature and the body of the relay. The gap between the core and the armature is filled with magnetic loops trying to contract. These contracting lines of force overcome the tension of the spring and pull the armature toward the core, operating the relay contacts. When the current in the coil is stopped, the magnetic circuit loses its magnetism and the spring pulls the armature up, opening the contacts.

Relays are useful in remote closing and opening of high voltage or high current circuits with relatively little voltage or current flow in the coil.

A relay can be used to switch an antenna between the transmitter and the receiver when the push-to-talk button is pressed on a microphone. Another example is the headlights or any other current drawing device in an automobile. The headlights need thick gauge wire to keep the resistance low. If we were to run heavy gauge wire to the switch of the dashboard it would require more wire and it would be difficult routing the wire and would have to connect it to a large heavy current switch.

Instead we run the heavy wire to the battery and relay contacts in series. The relay is very close to the battery so we do not need so much heavy gauge wire. Then we run small light wire to the dashboard to a lightweight low current switch. When we operate the switch to turn the lights on the relay operates and completes the circuit for the headlights.

Relay contacts are usually made of silver or tungsten. Silver oxidises but can be cleaned by using a very fine abrasive paper, or a piece of ordinary letterhead paper rubbed between the contacts. If the contacts are pitted by heavy currents, they may be smoothed with a fine file, but the original shape of the contacts should be retained to allow a wiping action during closing to keep them clean.

When I started out in electronics, relays were physically big components. These days relays for switching small currents can be made very small and look no larger than a medium sized integrated circuit.

The energy of a magnetic field.

Here is something for you to ponder. A magnetic field around a relay or a solenoid (electromagnet) has energy. The energy comes from the current flowing in the conductor coil. Suppose we switch the current OFF quickly by opening a switch. Where does the energy of the magnetic field go? It cannot just "disappear". Energy never disappears. It can change into another type of energy. Energy just can't go away.

Barkhausen Effect.

Earlier we explained how materials that can be magnetised contain many minute magnets in them called domains. For a long time, this was called the domain hypothesis of magnetism as it was not quite a fact. A scientist by the name of Barkhausen was experimenting with microphones attached to a soft iron bar while adjusting the strength of an external field. As he adjusted the strength of the external field, his microphone picked up the sound of the magnetic domains moving and the domain theory of magnetism was proven. He also noticed that the domains did not move smoothly. The domains moved in discrete steps. Staying still for a while as the magnetising force was increased; then suddenly jumping to their next position. This and other experiments led to the idea of what is known today as a quantum. A quantum (plural: quanta) is the minimum amount of any physical entity involved in an interaction. This became critical to the evolution of quantum mechanics.

9 - Alternating Current I

FARADAY'S LAW

When relative motion exists between a conductor and a magnetic field an emf is induced into the conductor.

This is a fundamental law. We will take our time and try to understand it thoroughly. What Faraday discovered was a way of producing electrical pressure (voltage or emf) from a magnetic field.

Faraday's law requires relative motion. This means there must be motion between the conductor and the magnetic field. This is not to say that the magnetic field must move. Neither does it mean that the conductor must move. Provided there is relative motion between the magnetic field and the conductor; a voltage will be induced into the conductor. In practice, this means the conductor can be stationary and the magnetic field must be moved relative to it. Alternatively, the magnetic field can be stationary and the conductor moved through it.

A Mind Experiment

Imagine a cardboard tube. I have successfully used the cardboard tube from an empty kitchen wrap. The tube is wound with at least 50 turns of insulated conductor. So, in fact, we have made a simple coil. To the ends of the coil, we attach a voltmeter. We have not discussed meters yet, but I think you can follow along if I just say it is an instrument with a needle, which indicates if the voltage is present. I used a special type of voltmeter called a galvanometer. A galvanometer is a voltmeter that has zero volts at centre scale. The advantage of this is that the meter's pointer can deflect left or right of centre depending on the polarity of the voltage being measured.

In figure 9-1 we have a galvanometer connected to our coil. Now imagine taking a permanent magnet, a bar magnet, and inserting it quickly into the centre of the coil.

The pointer will deflect in one direction as you insert the magnet. When the magnet stops moving the meter will return to zero.

You still have the magnet inside the coil. Now pull the magnet out. The pointer will again deflect, indicating a voltage on the coil, but, this time, the deflection will be in the opposite direction to when we inserted the magnet.

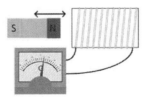

Figure 9-1

What we have done is demonstrated Faraday's law of induction. When we inserted the bar magnet inside our coil, we produced relative motion between the coiled conductor and the magnetic field of the bar magnet. Once the magnet was in and stopped moving, the relative motion stopped and the induced voltage fell to zero again. When we removed the magnet from the coil, relative motion existed again, but since the magnet was being extracted rather than inserted, the polarity of the induced voltage was in the opposite direction and the meter's pointer deflected accordingly.

If you happened to have made the electromagnet we discussed in an earlier chapter, you could place it inside the coil. If you then energised the electromagnet by applying a voltage to it, the expanding magnetic field would induce a voltage in our coil. After a few moments, the induced voltage would fall to zero. Now, if we disconnected the voltage to the electromagnet, the magnetic field around it would collapse (relative motion) and voltage would be induced in the coil. You may not know it, but you have just demonstrated the operation of a device to be discussed shortly, called a transformer.

THE AMOUNT OF INDUCED VOLTAGE

How much voltage is induced with our coil experiment is determined by the number of turns on the coil, the strength of the magnet we use and the speed of the relative motion.

Of course, if we were able to continue to move our bar magnet in and out of the coil at high speed without stopping, we would get a continuously induced voltage.

Remember how the needle of the galvanometer first defected one way and then the other? This is because when the magnet is inserted, the induced voltage is one polarity (negative or positive) and when the magnet is withdrawn the polarity reverses. This is an alternating voltage. An alternating voltage is one which changes polarity periodically. If we connect an alternating voltage to a circuit, we will get an alternating current. This is a current that flows in the circuit first in one direction and then in the other.

CYCLES PER SECOND

Going back to our coil. Placing the magnet into the coil and then removing it is one cycle. If we could take the magnet in and out of the coil every second, the polarity of the induced voltage would go through one cycle in one second. Mechanics call this revolutions per second. In electronics, we once used cycles per second, but now use Hertz. So when you see 'hertz' just think of 'cycles per second'.

This method of inducing an emf was the first method used for household power generation. The first household power was direct current (DC). The method used to produce this direct current was electromagnetic induction. The generator converted the AC to DC.

The alternating voltage connected to the household supply from the power grid changes polarity 100 times a second (50 cycles per second) and reaches a peak of 325 volts in Australia. The mains voltage in some locations varies a great deal depending on consumer load.

WHERE HAVE ALL THE ELECTRONS GONE?

A bit of a strange subtitle but it's all I could think of now. What I want you to think about is the electrons in your household wiring! They don't move from the power station through the conductors to your house. The household supply is AC at 50 hertz (Hz). So the electrons in the conductors in your house are probably the same electrons that were there when the wiring was installed in your home. The AC supply just causes them to move first one way and then the other. We discussed earlier how slowly electrons really move. So if they (the electrons) are changing direction 50 times a second, they would not be getting anywhere. They just see-saw back and forth. For the effects of an electric current to work, we don't have to make the electrons move in one direction all the time we just have to make them move. Think about that.

Let's look at how AC is produced in the power station. The device which creates the household power is called a generator. It works by creating relative motion between a conductor and a magnetic field.

Figure 9-2 is a simplified AC generator. The segments N and S are the north and south pole of a permanent magnet. Between the magnet, in the magnetic field, we have a coil of wire. Only one turn is shown but in practice, there would be many turns. The coil in a generator is called the armature. The ends of the coil are connected to a load via slip rings and brushes. We are going to make the armature rotate and we don't want the wires to become twisted where they connect to the load, so we feed them out through slip rings. We are going to rotate the armature through 360 degrees or one cycle and look what happens to the induced emf (voltage). When there is no relative motion between the armature and the magnetic field there will be no induced emf. When the armature is cutting across magnetic field line the relative motion will be the greatest and the induced emf will be at its greatest.

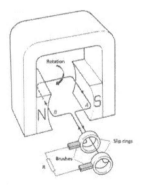

Figure 9-2

We can choose any starting position for the armature. In Figure 9-2 the armature is horizontal. We are going to start with the armature in the vertical position as shown in figure 9-3. In plan view we would see the armature as a straight line. In side view we would see all of the armature as shown. When the armature starts to rotate it will be at the instant rotation starts in line with the magnetic field and the induced voltage will be zero volts. As the armature continues to rotate it becomes more perpendicular to the magnetic lines of force the induced voltage increases rapidly, reaching a maximum when the armature reaches 90 degrees.

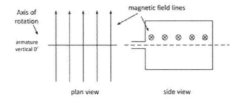

Figure 9-3

In figure 9-4 the armature has now reached 90 degree. Looking from above in plan view we see the entire armature which is now "in line" with the magnetic field line and has reached the peak value of the induced emf. The armature in 9-4 is in the same position as shown in figure 9-2.

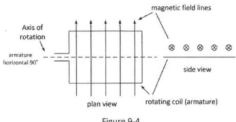

Figure 9-4

In figure 9-4 the relative motion is the greatest because the armature is cutting across the magnetic field line at a right angle to those field lines. Figure 9-5 shows the position of the armature from 0 to 360 degrees in steps of 90 degrees.

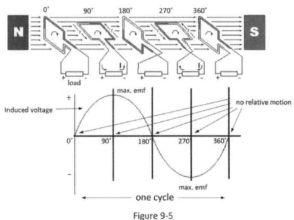

Figure 9-5

Figure 9-5 shows when the armature is cutting across field lines at 90 and 120 degrees the relative motion is the greatest and the induced emf is the greatest. At 0 and 270 degrees the armature is travelling along field lines and the relative motion is least as is the induced emf. You might also notice that the polarity of the emf changes each half cycle. We have alternating voltage and the shape of that voltage is a sinewave.

10 - Alternating Current II

In the last chapter, we discussed how an alternating current, or voltage, can be produced using a generator.

We are now going to look more closely at the shape (waveform) of this AC voltage. Recall how we discussed that the voltage starts out at zero, builds up to a maximum, then falls to zero. The polarity of the voltage then reverses. It builds up to a maximum again and then falls back to zero. This is one cycle. The number of cycles in 1 second is called the frequency and is measured in hertz.

Now the shape (waveform) of the wave from a generator is called a sinewave or sinusoidal wave. Some textbooks just flash this shape in front of you and say here it is. When I was learning I found the concept of a sinewave a little difficult, so I am going to assume the same for you.

A sinewave is the graphical representation of the alternating current or voltage plotted against time.

AN EXPERIMENT

Get yourself a piece of paper, about A4. On the left-hand side of the paper place a pen down and draw a vertical line about 50mm long. Now continue to trace over the line going up and down, over and over. Consider the centre of this line to represent zero volts. Why aren't you doing it?

Continue to move your pen up and down, redrawing the line. I want you to think that the middle of the line represents zero volts. Your pen moving upwards toward the top of the line represents current or voltage increasing in one direction. When you get to the top move down again and you eventually pass through the zero point. Once through zero your pen movement is now representing a change in the direction of the current and it begins to increase towards the bottom of the line, which represents maximum voltage in this direction. When you reach maximum at the bottom your pen now moves back toward the zero point and so on.

So you are sitting there copying over this vertical line. Do it reasonably quickly and get used to it.

Now without stopping your up and down motion, move your hand to the right without even thinking about the pen.

This is what I got, shown below in figure 10-1, though I have added a couple of notations.

The graph is very rough, but it does illustrate clearly the waveshape of an alternating voltage plotted against time. The waveshape is called a sinewave.

An AC voltage can be fed into an oscilloscope and its screen will display the waveform precisely.

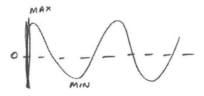

Figure 10-1

Figure 10-3 is how you will see the sinewave represented in most textbooks. Sometimes positive and negative symbols are drawn on the vertical axis. Just think of positive as being current flow in one direction and negative as current flow in the opposite direction. How many cycles you see will depend on the time scale on the horizontal axis. On an oscilloscope, the horizontal and vertical axes are calibrated accurately in time and voltage. So we can read things like the maximum voltage and we can work out the frequency in hertz (number of cycles per second).

The shape of the AC voltage delivered to your home is a sinewave.

One cycle, or revolution, is 360 degrees. Figure 10-2 shows the values of the voltage or current for any sinewave at different parts of the cycle. Sine is a trigonometric function that can be used to calculate points on a sinewave. You will find the 'sin' key on most calculators and if you enter 30 (degrees) into the calculator and press 'Sin' you will get 0.5. At 30° the instantaneous value of a sinewave is 0.5; with 1 being its peak value. On some calculators you press 'Sin' then enter 30° and '='. This is why the waveform is called a sinewave.

Degrees	Sin (degrees)	voltage
0	0	Zero
30	0.500	50% of maximum
45	0.707	70.7% of maximum
60	0.866	86.6% of maximum
90	1.000	Positive maximum
180	0	Zero
270	-1.000	Negative maximum
360	0	Zero

Figure 10-2

VOLTAGE AND CURRENT VALUES FOR A SINEWAVE

Suppose you wanted to buy electricity from Edison or Tesla. Edison says he will supply you with 100 volts direct current (DC), none of this sinewave stuff from him. He offers to provide 100 volts just like you would get from a battery. No current reversals. Tesla offers you 100 volts AC. From whom would you buy electricity?

Well, Edison's offer is very straightforward. Tesla, on the other hand, is offering a sinusoidal wave and you would have to ask more questions. Tesla is offering you 100 volts but what and when is it 100 volts? Is it 100 volts at the peak value of each half cycle? Which would mean that you would only get 100 volts for a brief moment twice during each cycle. Does this matter? You bet it does. If you were in the middle of winter in front of the electric bar heater, you would want 100 volts all the time not just for a brief period, each half cycle.

65

100 volts peak AC will not produce the same heating effect as 100 volts DC. So there is a need to be able to compare AC with DC, in terms of their heating effect. That is the same as comparing how much work each can produce.

It can be found mathematically or by experiment that the equivalent heating effect of a sinusoidal waveform compared to DC is 0.707 of its peak value. This value is known as the root mean square or effective value and is written RMS.

So if Tesla was to supply 100 volts peak, the actual equivalent heating value would be:

0.707 x 100 = 70.7 volts RMS, in which case you would go for Edison's deal? The average value of a sinewave is:

Peak x 0.637

I have never found a use for this value.

Sometimes an AC voltage is given as peak-to-peak. So if the maximum peak voltage was 100 volts, then the peak to peak value would be 200 volts - again a need to know however, I have never found any use for the peak to peak value.

Your household electricity supply is 240 volts RMS. This means it provides the same heating value as 240 volts DC. What is the peak value of the household power supply in Australia? The Standard for AC mains in Australia is now 230V AC RMS. However, it is nearly always closer to 240V AC RMS.

RMS = 0. 707 x Peak

Transpose for peak by dividing both sides by 0.707 Peak= RMS/0.707 = 240/0.707 = 339.46 volts

So twice during each cycle, the household mains voltage actually reaches 339 volts. The frequency of the household supply is 50 hertz. How many times a second does the mains voltage reach 339 volts?

Since there are 50 cycles per second and two peaks during each cycle, the supply voltage will reach 339 volts 100 times each and every second.

Summary of equations

RMS = 0. 707 x Peak
Peak= RMS/ 0.707
Average= Peak x 0.637
Peak-to-Peak = Peak x 2

Note: 1/0.707 is 1.414 and 1.414 is the square root of 2 The duration of one cycle in seconds is called the period. What is the period of the household mains in Australia?

Since the frequency is 50Hz, there are 50 cycles in 1 second. Therefore 1 cycle has a Period of 1/50th of a second or 20 milliseconds.

Period = 1 / frequency.

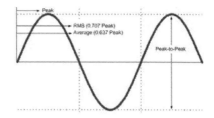

Figure 10-3

WHAT DOES ROOT MEAN SQUARE MEAN?

We have learned what the RMS value of an AC voltage means. We also know the RMS value is 0.707 of the peak. However, what is this term 'root mean square'?

The factor 0.707 for RMS is derived as the square root of the average (mean) of all the squares of a sinewave. Now, if that still sounds like gobbledygook, let me show you how it is done and at least we will take away the mystery of the term RMS.

Interval	Angles	Sin(angle)	Sin(angle)^2
1	15	0.26	0.07
2	30	0.50	0.25
3	45	0.71	0.50
4	60	0.87	0.75
5	75	0.97	0.93
6	90	1.00	1.00
7	105	0.97	0.93
8	120	0.87	0.75
9	135	0.71	0.50
10	150	0.50	0.25
11	165	0.26	0.07
12	180	0	0.00
	Total	7.62	6.00
	Average	7.62/12	SQRT(6/12)
		=	=
		0.635	0.707

Figure 10-4

Michael Faraday, who became one of the greatest scientists of the 19th century, began his career as a chemist. He wrote a manual of practical chemistry that revealed his mastery of the technical aspects of his art. He discovered a number of new organic compounds, among them benzene and was the first to liquefy a "permanent" gas (i.e., one that was believed to be incapable of liquefaction). His major contribution, however, was in the field of electricity and magnetism. He was the first to produce an electric current from a magnetic field; invented the first electric motor and dynamo; demonstrated the relationship between electricity and chemical bonding; discovered the effects of magnetism on light; and discovered and named diamagnetism, the peculiar behaviour of certain substances in strong magnetic fields. He provided the experimental and a good deal of the theoretical, foundation upon which James Clerk Maxwell worked out the classical electromagnetic field theory.

67

11 - Capacitance - I

In this chapter, we are going to talk about capacitance. We need to make a distinction between the capacitor and capacitance. A capacitor is a device, whereas capacitance is an electrical property. First, we will discuss the capacitor and then the property of capacitance.

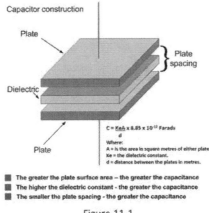

Figure 11-1

As you can see a capacitor is a two-terminal device. Two conductive plates separated by an insulator. This should suggest to you that current never flows through a capacitor.

Capacitor types generally take their name from the type on insulation material used between the plates. This is called the dielectric of the capacitor. Some capacitors have their dielectrics made electrically. These capacitors are said to be polarised in that they have a positive and negative terminal. When connected in a circuit this polarity must be maintained, if not the dielectric will be destroyed and the capacitor could gas and explode.

Figure 11-2 shows the standard schematic symbols for a capacitor.

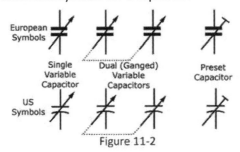

Figure 11-2

Some capacitor types and packages: -

Figure 11-3

Figure 11-3 shows a variety of capacitor types. The yellow/brown are ceramic capacitors (because they use ceramic as the dielectric). The green capacitor in the foreground is the polyester type. The blue and black cylindrical capacitors are called electrolytic. The pale blue bottom left is a tantalum capacitor. Electrolytic and tantalum capacitors have their dielectrics formed electrically, for this reason, they are polarised capacitors that have a negative and positive leg. Connected incorrectly they can create internal gas and explode.

Larger 200-400pF air variable capacitors are used in transmitter output stages and antenna tuners. Air dielectric capacitors can withstand higher voltages. It takes about 10kV to flash over 1cm of dry air. Notice that in addition to the capacitance in farads, capacitors also have a voltage rating.

Figure 11-4

Charging and discharging a capacitor

The interesting stuff starts to happen when we connect an emf to the plates of a capacitor. Have a look at the test circuit of figure 11-5.

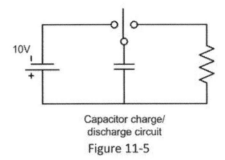

Capacitor charge/
discharge circuit

Figure 11-5

The capacitor can be switched so that it will be connected to the 10 volt battery when the switch is thrown to the left, or when the switch is thrown to the right, it will be connected to the resistor. The capacitor can be connected by the switch to the battery or the resistor, but not both at the same time.

The negative terminal of the battery has an excess of electrons on it, created by the chemical action within the battery. The positive terminal has a deficiency of electrons. Now recall that unlike charges attract. If there was any way for the electrons to get from the negative terminal of the battery to the positive, they would.

Now I just want you to imagine a battery by itself with two terminals. There is an electrostatic field between the two terminals of any battery created by the unlike charges on each terminal. In other words, there is a very slight tugging from the positive terminal and a very slight pushing from the negative terminal

in a vain attempt to move electrons from the negative to the positive terminal. No current flows between the unconnected terminals of a battery because the electrostatic fields are very small due to the spacing of the battery terminals and the very high air resistance between them.

If you find this hard to imagine look at figure 11-6 of two charges separated by air.

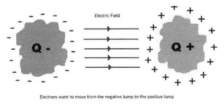

Figure 11-6

Here we have two lumped charges Q- and Q+. There is an electrostatic field between them. This field is trying to pull electrons from the negative charge over to the positive charge. There is an electrical strain here. No current flows because the two charges are too far apart. Let's say they are 100mm to start with. Now move the two lumped charges so that they are only 50mm apart. The electrostatic field will now be stronger. The strain will be stronger. Still no current flows. Let's push the issue. Move the two lumped charges so that they are only 5mm apart. Now (depending on the charges) the electrostatic field will be much stronger. The tugging of electrons from the negative charge will be much greater. Still no current flows, as the air insulation resistance between the charges is too high.

The electrons on the negative lumped charge want to traverse the gap to the positive charge. Do you think the electrons would be evenly distributed on the negative lump? On the side of the negative lump closest to the positive lump (the inside) the electrons will be crowding up trying to jump the gap.

Can you figure out what will happen if we continue to move them closer, say, to 1mm? I think you will agree that a point will be reached where the electrostatic field is so strong that electrons will jump off the negative lump and flow through the air to the positive lump. There will be an arcing of electric current. This is what happens in a lightning storm.

Returning to our capacitor that, at the moment, is not connected to anything. The switch is thrown to the left. The plates of the capacitor have a very large surface area.

The dielectric between the plates is extremely thin but still a very good insulator. When we throw the switch to the left as in figure 11-7, we are extending the charges on the battery terminals to the plates of the capacitor and there will be a strong electrostatic field across the plates of the capacitor and through the dielectric.

Electrons are going to move from the negative terminal of the battery and bunch up on the top plate of the capacitor. Similarly, electrons are going to move from the bottom plate and travel to the positive terminal of the battery. No electrons will be able to flow through the dielectric. Its insulating properties are too good. So, if you like, there is going to be a redistribution of charge. This movement of charge will continue until the electrons flowing into and out of the capacitor create a potential difference on the plates of the capacitor equal to the battery voltage, namely 10 volts. The capacitor is now said to be charged.

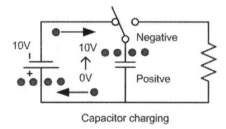

Capacitor charging

Figure 11-7

This charging of the capacitor does not happen instantaneously - it takes a little time. Suppose now we move the switch so the capacitor is disconnected from the circuit, as shown in figure 11-8.

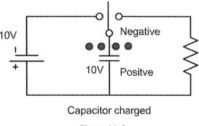

Capacitor charged

Figure 11-8

The capacitor is now left charged, even though it has been disconnected from the battery. The capacitor has 10 volts across it. Older capacitors would not hold this charge for very long as a current would slowly leak through the dielectric and the capacitor would eventually self-discharge. Some modern capacitors can hold their charge for many days or even weeks.

The capacitor has stored energy due to the charge on the plates. If we connect a circuit to the capacitor, it will discharge as shown in figure 11-9.

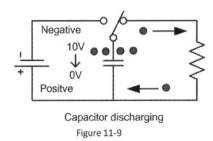

Capacitor discharging
Figure 11-9

Electrons will now flow from the top plate of the capacitor through the resistance until the capacitor becomes discharged. If the resistor was a small light bulb, it would flash brightly at first and then slowly dim as the capacitor discharges.

At no time did current flow through the dielectric of the capacitor. Current flows into and out of a capacitor giving the illusion that current flows through a capacitor.

Suppose the resistor was a lamp. Also, suppose we continue to move the switch rapidly back and forth between the left and right position. The lamp would perhaps flicker a bit, but continuously give off light.

So you should now see that a capacitor can be used to store charge and we can use that charge to do something.

Where is the energy stored?

Though we have said that energy has been stored due to the charge on the plates, it is more correct to say that the energy is stored in the electric field. It is the charge on the plates that forms the electric field between the plates. When current flows into a capacitor, charging it, the electric field becomes stronger (stores more energy). When the current flows out of the capacitor, the voltage across the plates decreases and hence the strength of the electric field decreases (energy moves out of the electric field).

UNIT OF CAPACITANCE

The unit of capacitance is called the Farad. The farad is the measure of a capacitor's ability to store a charge. If one volt is applied to the plates of a capacitor and this causes a charge of 1 coulomb to be stored on the plates, the capacitance is 1 farad.

In practice 1 farad is an enormous capacitance. More practical sub-units of the farad are used. Microfarad and picofarad are the most common sub-units.

1 microfarad = 1 x 10^{-6} farads
1 nanofarad = 1 x 10^{-9} farads
1 picofarads = 1 x 10^{-12} farads

Capacitors range from as small as 1 picofarad to as high as thousands of farads. The very high value capacitors are electrolytic and have a working voltage of about 2.5 to 3V. The very high capacity low voltage capacitors are called "Super Capacitors".

A 2.7V 3000F Super Capacitor Figure 11-10

PERMITTIVITY OR DIELECTRIC CONSTANT

The insulating material between the plates (dielectric) determines the concentration of electric lines of force. Just like different materials will concentrate magnetic lines of force to a greater or lesser extent, materials also vary in their ability to concentrate electric lines of force.

If the dielectric was air, then a certain number of lines of force will be set up. Some papers have a dielectric constant twice that of air, which would cause the density of the electric lines of force to be double and the capacitance would be double. The ability of a dielectric to concentrate electric lines of force is called the dielectric constant or permittivity. The higher the dielectric constant, the greater the capacitance for a given plate area.

Suppose an air dielectric capacitor (dielectric constant close enough to 1) of 8 microfarads had its air dielectric replaced with mica, without changing the distance between the plates. The capacitance would increase in direct proportion to the dielectric constant.

In other words, the capacitance would increase from 8 microfarads to 5-7 times that value, or 40 to 56 microfarads proportional to the new dielectric constant.

DIELECTRIC CONSTANTS

Material	Dielectric Constant
Vacuum	1
Air	1.0006
Rubber	2-3
Paper	2-3
Ceramics	3-7
Glass	4-7
Quartz	4
Mica	5-7
Water	80
Barium titanate	7,500

Figure 11-11

FACTORS DETERMINING CAPACITANCE

A formula to determine the capacitance of a two-plate capacitor is:

$$C = \frac{KeA}{d} \times 8.85 \times 10^{-12} \text{ F}$$

Where:

A = is the area in square metres of either plate.

Ke = the dielectric constant.

d = distance between the plates in metres.

The constant 8.85×10^{-12} is the absolute permittivity of free space.

Example

Calculate C for two plates, each with an area of 2 square metres, separated by 1 centimetre (1×10^{-2} metres), with a dielectric of air.

We will take the dielectric constant of air as 1. Even though it is more accurately 1.0006, 1 is close enough, so that:

$$C = \frac{1 \times 2 \times 8.85 \times 10^{-12}}{1 \times 10^{-2}}$$
$$= 200 \times 8.85 \times 10^{-12}$$
$$= 1770 \times 10^{-12}$$
$$= 1770 \text{ pF (picofarads)}$$

For examination purposes, you do not have to use this equation. However, you most definitely do need to know what the equation says about the factors determining capacitance.

$$C = \frac{KeA}{d} \times 8.85 \times 10^{-12} \text{ F}$$

Capacitance is directly proportional to the dielectric constant (Ke). Capacitance is directly proportional to the area of one of the plates (A). Capacitance is inversely proportional to the distance between the plates (d).

CAPACITORS IN SERIES AND PARALLEL

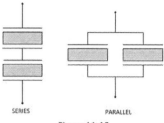

SERIES   PARALLEL

Figure 11-12

In figure 11-12 all the capacitors are the same. Let's say they are 10 microfarads. Think about what happens when you connect two identical capacitors in series. Think about it in terms of the factors that we just discussed; that affect capacitance. Can you visualise by looking at the series capacitors that we have actually doubled the thickness of the dielectric? Doubling the thickness of the dielectric is exactly the same thing as doubling the distance between the plates.

If the distance between the plates is doubled and capacitance is inversely proportional to the distance between the plates, then the capacitance must be half that of a single capacitor on its own. The total capacitance of two 10 microfarad capacitors in series must then be 5µF.

The equation for any number of capacitors of any value in series is:

$$1/C_t = 1/C_1 + 1/C_2 + 1/C_3 + ..n$$

OR

$$C_t = \frac{1}{1/C_1 + 1/C_2 + 1/C_3 + ..n}$$

The easiest way to use this equation on a calculator is:

1. Find the reciprocal of each capacitance.
2. Add all of the reciprocals together (sum them).
3. Find the reciprocal of this sum.
In other words, the reciprocal of the sum of the reciprocals.

Worked example

Take three capacitors of 7, 8 and 12 microfarads in series.

Find the reciprocal of 7: 1/7 = 0.143 (rounded to 3 decimal places)
The reciprocal of 8: 1/8 = 0.125
The reciprocal of 12: 1/12 = 0.083
The sum of the above is: 0.351
Find the reciprocal of this sum: 1/0.351 = 2.85 microfarads.
Three capacitors of 7, 8 and 12 microfarads in series have a total capacitance of 2.85µF.
Notice how the total capacitance is always less than the lowest value capacitor.

If just two capacitors are in series then you can use the simplified product over sum formula.
For two capacitors (only) in series:

Ct = (C1 x C2) / (C1 + C2)

When the same two capacitors are connected in parallel, the distance between the plates and all other factors remain the same, except we have doubled the effective area of the plates. So the capacitance has doubled.

The equation for any number of capacitors of any value in parallel is:

Ct= C1 + C2 + C3 + ... n

These equations are opposite to that of resistances in parallel and series, so be careful not to confuse the two. The shortcuts we took with resistors in parallel work the same for capacitors in series. For example, if we have just two capacitors of any value in series we can use the product over sum method to find the total capacitance.

VOLTAGE ACROSS SERIES CAPACITORS

If two equal capacitors were connected in series across a 100 volt DC supply and we were to measure the voltage across each capacitor, we would get 50 volts across each. Since the capacitors are equal, we get an equal voltage drop.

For unequal capacitances in series the voltage across each C is inversely proportional to its capacitance. In other words, the smallest capacitance would have the largest voltage drop. The largest capacitance would have the smallest voltage drop.

The amount of charge on each series capacitor is given by:

Q=CE

If a 10μF capacitor was charged to 10 volts, then the charge in coulombs on the capacitors would be:

10μF x 10 volts= (10×10^{-6})F x (10)V = 1×10^{-4} coulombs. (100μC)

The equation can be transposed for voltage across a capacitor and we get:

E =Q/C

In a series circuit, each capacitor regardless of its capacitance will have the same charge. Q is the same for all capacitances in series. For a smaller capacitance to have the same charge as larger capacitors in a series circuit, it must have a higher E.

We could do numerical examples. However, you do not need this in practice or for examination purposes. You do need to understand and be able to visualise the voltage drops across capacitances in a series circuit.

Example.

Two capacitances of 1μF and 2μF are connected in series across a 900 volt DC supply. What is the voltage drop across each capacitor?

Now, the voltage drops have to be unequal because each capacitor will have the same charge (Q).

E=Q/C

E is inversely proportional to C. Therefore, the smallest C (1μF) must have the greatest voltage drop. But how much greater? Since the 1μF capacitor is half the value of the 2μF capacitor, it must have twice the voltage to achieve the same charge. The 1μF capacitor must then have 600 volts across it, leaving 300 volts on the 2μF capacitor.

VOLTAGE RATING OF CAPACITORS

All capacitors are given a maximum voltage rating. This is necessary as the dielectric of capacitors can breakdown and conduct causing the capacitor to fail and in most cases be destroyed. Some capacitors, if placed across a voltage which is too high, will create gas within them and explode somewhat violently.

Thomas Alva Edison was the quintessential American inventor in the era of Yankee ingenuity. He began his career in 1863, in the adolescence of the telegraph industry, when virtually the only source of electricity was primitive batteries putting out a low voltage.

Before he died in 1931, he had played a critical role in introducing the modern age of electricity. From his laboratories and workshops emanated the phonograph, the carbon-button transmitter for the

telephone speaker and microphone, the incandescent lamp, a revolutionary generator of unprecedented efficiency, the first commercial electric light and power system, an experimental electric railroad and key elements of motion-picture apparatus, as well as a host of other inventions.

Singularly or jointly he held a world-record 1,093 patents. In addition, he created the world's first industrial research laboratory.

Born in Milan, Ohio, on Feb. 11, 1847, Edison was the seventh and last of four surviving children of Samuel Edison Jr. and Nancy Elliott Edison. At an early age, he developed hearing problems, which have been variously attributed but were most likely due to a family tendency to mastoiditis. Whatever the cause, Edison's deafness strongly influenced his behaviour and career, providing the motivation for many of his inventions.

12 - Capacitance - II

TYPES OF CAPACITORS

AIR DIELECTRIC CAPACITORS

Capacitors using an air dielectric are used mostly as variable capacitors. Air dielectric variable capacitors are large but can withstand high voltages; this is just what is needed for transmitter output stages and antenna tuning units.

We have seen that capacitance is proportional to the plate area and inversely proportional to the spacing. Even with close spacing, the area of the plate must be large, in order to obtain the required capacitance. Instead of using two large plates, which would be inconvenient, a number of smaller plates can be interleaved.

The capacitance can easily be varied by sliding the plates in and out of mesh since the capacitance is proportional to the cross section of the dielectric between the plates connected to opposite terminals.

A practical capacitor is constructed with a set of fixed plates and a set of moving plates that rotate on a spindle. As the moving plates are rotated through 180 degrees the meshing and therefore the capacitance varies from a minimum to a maximum value. The maximum value of the capacitance is generally about 500pF. It is common practice to use two or three of these capacitors ganged together on a common spindle to form the main tuning control of a transmitter.

Figure 12-1

Over the years, the physical size of variable capacitors has decreased considerably due to the use of smaller spacing between the moving and fixed plates. However, with the introduction of small communication receivers; air spaced capacitors are too large. In receivers, a synthetic dielectric is now used. In these capacitors, a solid dielectric in the form of thin plastic sheets is placed between the plates. In this way, the capacitance is increased by the value of the relative permittivity of the dielectric. The sheets are a fairly loose fit between the plates so that the moving plates can still slide over the plastic sheets (which are fixed).

Figure 12-2

SILVER MICA CAPACITORS

This capacitor consists of thin mica sheets as the dielectric. The mica sheets are coated with a thin layer of silver, which forms the plates. A number of plates may be stacked together to obtain the required capacitance. The capacitor is protected by a wax, lacquer or plastic coating. These capacitors are available up to 10,000 pF and have a small loss, small tolerance (e.g. ± 1 %), are stable in capacitance value and have a low temperature coefficient. They are mainly used where an accurate, highly stable, good quality capacitor (high Q) is required. Typical applications of a mica capacitor include the crystal oscillators and other critical RF circuits.

There was a time not that long ago when you could identify most capacitors from the colour of the package and their general appearance. That is not always the case now. You may need to look up a manufacturer's code to identify a capacitor type. Mica capacitors are generally dark or light brown. Do watch out as these conventions may no longer apply. The package can be any shape and colour. An example of mica capacitors is shown in figure 12-2.

CERAMIC CAPACITORS

These capacitors use a ceramic for the dielectric and are very widely used as they are physically small and inexpensive. The term 'ceramic' covers a very large range of materials and the properties of the capacitor depend on the type of ceramic used.

The permittivity of some ceramic materials is very high, as much as 16,000; this results in a physically small capacitor. Capacitors can be made using relatively low permittivity ceramics, which have different temperature coefficients and can be used for temperature compensation. When a high permittivity ceramic is used the temperature coefficient may be very large and not linear, i.e. varies greatly with temperature. These also have a capacitance that changes with the applied voltage and because of this, the tolerance may be large, as much as 20% to +80%. Obviously, such capacitors must only be used where a precise capacitance value is not important.

The maximum working voltage may range from 100V to 10kV or more.

Ceramic capacitors are widely used owing to their small size, low cost and because so many types are available.

There are many bypass and coupling capacitors (discussed later) in a transceiver. There could be thirty or more. These bypass and coupling capacitors then need to be small, inexpensive and stable. This is an ideal application for the ceramic capacitor. Remember that tip.

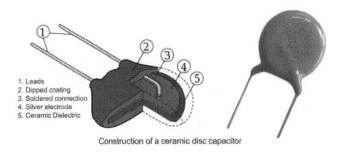

Construction of a ceramic disc capacitor

Figure 12-3

PAPER CAPACITORS

At one time paper capacitors were very common. They consisted of a metal foil for the electrodes and paper (impregnated with oil or wax) for the dielectric. Instead of a number of plates, two plates only are used, about 2 to 5 cm wide and of a length corresponding to the capacitance required. The plates and dielectric are now wound into a roll to form a tubular capacitor. Paper capacitors are rarely seen today.

PLASTIC FILM CAPACITORS

These are constructed similar to a paper capacitor, but with a plastic film instead of paper. There are several types: polystyrene, polycarbonate, polyester, polyethylene terephthalate or polypropylene. The plates may be foil (e.g. with polystyrene) or metallised plastic film. Polystyrene ones do not normally have a case, but the others are usually enclosed in a plastic case that may be cylindrical or rectangular. The various films have different detailed characteristics but will be considered as one class.

ELECTROLYTIC CAPACITORS

This type is used where a high value is required in a small space. It depends on the principle of depositing, by electrolytic action, an extremely thin insulating film on one plate, the film then acting as the dielectric. With such a thin dielectric, the capacitance for a given plate area is large. There are two basic types: aluminium and tantalum. These capacitors are polarised. If connected to the wrong polarity they will be destroyed.

ALUMINIUM CAPACITORS

These are made in a similar way to a paper capacitor but use two aluminium plates and an absorbent paper. The paper is impregnated with a suitable electrolyte, which produces an aluminium oxide coating on the positive plate. The paper acts as a conductor and container of the electrolyte, the only dielectric being the oxide coating. The capacitance may be increased by etching the positive plate (the one with the oxide coating), thus increasing the effective surface area. In normal capacitors, the aluminium coating must be maintained by the application of a direct voltage across the plates in the correct direction, i.e. the capacitor is polarised. If the voltage is reversed the coating will be removed and the capacitor ruined. Working voltages vary from 3 to 500 volts and it is important that the working voltage is not exceeded as they may explode. The capacitor passes a small leakage current, particularly when first switched on, reforming the oxide layer. This current rises rapidly if the working voltage is exceeded and damage then results. The capacitors are usually fitted into aluminium cases, the case often being one of the connections. These capacitors are commonly used in transistor equipment. Values are available from 1μF to 100,000μF. The tolerance is high, 25% to 100% and the losses are high, particularly at high frequencies.

There are some special electrolytic capacitors made that are not polarised, in which both plates have oxide coatings. They are used in a few special applications where a large electrolytic capacitor is require and polarity can change, e.g. crossover networks in loudspeakers.

Figure 12-4

Packages differ a great deal; from the small ones shown in 12-4 to large cylindrical packages. A typical application is coupling and bypassing in low frequency circuits to large filter capacitors found in low voltage DC power supplies.

When electrolytic capacitors fail they often bulge due to the internal generation of gas. When looking for a fault check for this bulging. If you find two or three electrolytic capacitors like this on the one circuit board it is good engineering practice to replace all of the electrolytics on that board.

TANTALUM CAPACITORS

The basic principles are the same as aluminium, with tantalum in place of aluminium. They have a better shelf life and lower leakage current but are more expensive. There are three types: wet electrolyte sintered anode, foil electrode wet electrolyte and solid electrolyte sintered anode. The last type is the cheapest and the more commonly used in domestic equipment, so no further details will be given of the first two types. The solid type uses a sintered tantalum anode with a solid electrolyte of manganese dioxide. It comes in various forms, in particular, tubular and "tear drop" types. Capacitance values vary from 0.1 to at least 100µF with working voltages up to 50V.

TIME CONSTANT

If a capacitor 'C' is connected to a DC supply through a resistor 'R', as shown in figure 12-5, a current will flow through R until the capacitor is charged to the supply voltage. As the capacitor charges, the voltage across it will rise to 63.2% of its final value, E, in a time equal to CR seconds (where C is in farads and R is in ohms). This time is known as the *time constant* of the circuit. The voltage across R is the difference between the supply voltage and the voltage across the capacitor and therefore decreases as the capacitor charges. As the capacitor charges the current will decrease and fall to zero at full charge.

If the supply is now disconnected the capacitor will remain charged, apart from the small leakage through the dielectric of the capacitor. It may remain nearly fully charged for hours or even days if it is a high Q (low loss) capacitor.

A large capacitor (e.g. 1 Farad) charged to a high voltage (say 500 volts) may store considerable energy and it is unwise to touch the terminals or a nasty shock will result. If the resistor is connected across the charged

81

capacitor, it will discharge by passing a current through R, the value of the current being determined by Ohm's law. The voltage will now decrease by 63.2% in a time equal to the time constant 'CR'.

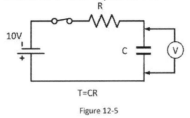

T=CR

Figure 12-5

Example

In figure 12-5, the resistance is 1MΩ and the capacitor 10μF. What is the time constant of the circuit?

T=CR = $10 \times 10^{-6} \times 1 \times 10^{6}$ = 10 seconds.

When the switch is closed, what will be the voltage on the charging capacitor after 10 seconds?

10 seconds is 1 time constant period. The capacitor will charge to approximately 63.2% of the applied voltage after 1 time constant period has elapsed. Hence, after 10 seconds, the voltage on the capacitor will be about 63.2% of 10 volts or 6.32 volts.

I would like to pause here for a moment to emphasise something. Consider a single resistor connected across a battery with a switch and the switch was closed. How long would it take for the supply voltage to appear across the resistor? Answer - immediately. With a capacitive and resistive circuit, however, the voltage is delayed from building up on the capacitor, determined by the RC time constant. This gives us a better definition of capacitance.

Capacitance is that property of an electric circuit, which opposes changes of voltage.

Going back to our circuit in figure 12-5. The capacitor has charged to 6.32 volts after 1 time constant period. How much further does it have to charge before it equals the supply voltage? The answer is: 10 - 6.32 = 3.68 volts.

During the next time constant period (10 seconds) the capacitor will charge a further 63.2% of the remaining voltage. That is, 63.2% of 3.68 volts, which is 2.325 volts. So after 20 seconds the capacitor will have charged to about 6.32 + 2.325 = 8.645 volts.

In other words, the capacitor charges to 6.32 volts in one time constant period (10 Sec) and it charges a further 2.325 volts in the second time constant period. So, after 20 seconds, the total voltage on the capacitor will be 8.645 volts. During the third time constant period the capacitor will again charge a further 63.2% of the remaining voltage and so on. A capacitor can be considered to be fully charged after 5 time constants.

The same time constant applies to discharging the capacitor through 'R". If the capacitor is charged to 10V then discharged through a resistor after one time constant it will discharge by 63.2% of the starting voltage of 10V. Then in the second time constant period discharge a further 63.2% of the remaining voltage. After 5 time constants the capacitor is considered to be discharged.

Using equation T=CR allows us to easily work the voltage on a charging or discharging capacitor. However you can only do it in increments of a time constant. It is also obvious that going by time constants the capacitor is never (at least mathematically) fully charged.

Figure 12-6 shows the charging voltage across a capacitor and the current through it. The top of the figure shows the voltage increasing on the capacitor over 5 time constants. Notice that at the instant the switch is closed the voltage on the capacitor is zero volts. Then after the first time constant the capacitor charges by 63.2% or to 0.63V of what will then be the final voltage of 1 V.

The bottom part of figure 12-6 shows the current flowing in to the capacitor over 5 time constants. Notice that at the instant the switch is closed; when time equals zero; the current into the capacitor is at maximum. The current slowly decrease and stops after 5 time constants.

THE CONCEPT OF LEADING CURRENT

Compare the current and voltage in figure 12-6 to that of a resistive circuit. In a resistive circuit when the switch is closed the voltage across and the current through a resistance reaches the Ohm's Law value immediately.

If you were describing current and voltage in a capacitive circuit to someone you would have to say the current starts high and tapers off to nothing while the voltage starts at zero and increases to a maximum.

So current and voltage in a capacitive circuit are not in sync. Current is said to lead the voltage in a capacitive circuit. We describe this behaviour by saying: -

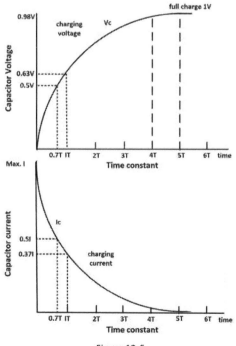

Figure 12-6

THE VOLTAGE AFTER ANY TIME -T

We can find out the voltage on a charging or discharging capacitor in increments of a time constant by using the equation T=CR. Suppose we wanted to know the voltage after any time period. Take a look at figure 12-7.

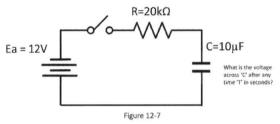

Figure 12-7

The time constant of this circuit is T=CR = 20kΩ x 10μF = 0.2 seconds.

Suppose we wanted to know the voltage across the capacitor 0.5 seconds after the switched was closed. We can use the equation below. We are able to do so because the charge and discharge curve is described by the natural logarithm.

$$Ec = Ea(1 - e^{-(\frac{t}{RC})})$$

Where
Ec = the voltage across the capacitor after time 't'
Ea = the applied voltage
e = the base of the natural logarithm
R = resistance in Ohms
C = Capacitance in Farads

$$Ec = Ea(1 - e^{-(\frac{0.05}{0.2})})$$
$$Ec = 11.01498002 \; volts$$

The voltage across the capacitor 0.5 seconds after the switch is closed is 11 V.

RC time constants have many uses in radiocommunications. The de-emphasis and pre-emphasis networks in FM are produced by a simple RC time constant circuit. RC time constants are used for time delays, simple high and lowpass filters and key click filters.

Analogy of a charging capacitor

Suppose you were to blow up a balloon. Some balloons have an initial resistance to being inflated so let's take one that has already been blown up and deflated. The flow of your breath going in the balloon is current. The pressure you create with your breath is voltage. When you first start to inflate the balloon, it is easy and there is a high inrush of air (current).

So air flow or current is high at first. After just one breath the balloon has pressure pushing back against you. After one breath the balloon does not have much back pressure (voltage).

As you continue to inflate the balloon, the backward pressure from it becomes greater. It becomes harder to inflate further. The amount of flow (current) into the balloon is less due to the amount of pressure (voltage) that the balloon is pushing back.

When the back pressure of the balloon matches your maximum inflating forward pressure - that is it - you cannot inflate the balloon further and the flow (current) comes to a stop. The balloon is now at maximum inflation, assuming it does not burst.

Can you see this is exactly how a capacitor behaves when it is being charged?

STRAY CAPACITANCE

When we build a capacitor, we want capacitance. However, all we need for capacitance are any two conductors separated by an insulator (dielectric). In any circuit, there are literally hundreds of stray and sometimes unwanted capacitances. More often these are not a problem, however, sometimes they can be. We will discuss this in more detail later.

A lifetime ago I was teaching PMG (telephone) linesmen about fixing telephone lines. A typical telephone line is just two insulated conductors twisted together and they can go for many kilometres.

A telephone line then by its very construction is a capacitor. Suppose a telephone line is only a kilometre long. Also nothing is connected to the telephone line - no telephone and no exchange equipment. The line is said to be open circuit. When a linesman connects an ohmmeter to the line an infinite resistance is expected. This is what an ohmmeter does on a good telephone line, but not immediately. The ohmmeter will swing across and to the right and slowly fall back to infinite ohms after one to two seconds. The ohmmeter shows the capacitance of the line charging.

An ohmmeter is a battery, resistor and a current meter connected in series. When the ohmmeter is attached to the line it shows at first a low resistance as the line acting as a capacitor and charging and after five time constants it is fully charged and the ohm meter reads open circuit. This technique can be used to estimate the length of the line.

13 - Capacitive Reactance

CAPACITOR IN A DC CIRCUIT

We have already discussed the operation of a capacitor in a DC circuit, let's just go over the main principles again.

If a capacitor is connected to a battery (or other DC source), it will charge according to its time constant (T=CR), to the battery voltage.

If a lamp is connected in series with the capacitor while it is charging, the lamp would give off light indicating that current is flowing. The lamp would be bright at first and then slowly dim and extinguish as the capacitor becomes charged. Current into the charging capacitor is high at first and then tapers off to zero when it is fully charged. The voltage on the capacitor is low at first and increases to the supply voltage when the capacitor is fully charged. We can see that the voltage across and the current into a capacitor, are not in sync (phase). Current in a purely capacitive circuit leads the voltage (by 90 degrees).

Imagine a capacitor and a lamp in series. Connect the circuit to a DC source and the lamp will light up momentarily. The capacitor is now charged and the lamp is extinguished. Suppose now we reverse the battery (or supply terminals). The capacitor will now discharge through the lamp and the battery and then recharge with the opposite polarity.

Imagine now continuing to do this very quickly, that is, reversing the battery every time the capacitor is charged or near to fully charged. Because the capacitor will be continually charging and discharging, there will be current flow in the circuit all the time and the lamp will be continuously lit.

Instead of reversing the battery, it would be much easier to supply the circuit with AC voltage, which by definition, automatically changes polarity each half cycle.

CAPACITOR IN AN AC CIRCUIT

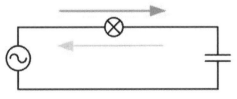

Figure 13-1

In the circuit of figure 13-1, a capacitor and a lamp are connected via a switch to a source of AC. The capacitor will continually charge and discharge as the AC supply changes polarity. AC current will flow in the circuit continuously and the lamp will remain lit. The two arrows indicate the charge and discharge currents.

So a capacitor 'appears' to allow an AC current to pass through it. I say 'appears' because while there is current flowing in the circuit at all times, at no point does current actually flow through the capacitor. Remember it has a dielectric which will not pass current. The current flows continuously in the circuit because the capacitor is constantly charging and discharging. This type of current is referred to as a displacement current.

Capacitance is the property of a circuit that opposes changes of voltage. A capacitor, therefore, has an opposition to current flow. The opposition to current flow produced by a capacitor is called Capacitive Reactance and is measured in ohms. The shorthand for capacitive reactance is X_C.

FACTORS DETERMINING X_C

What do you think we could do in the above circuit to increase the brightness of the lamp without changing the lamp or the supply voltage?

Firstly, let's reflect on the frequency of the AC supply. It helps to go to extremes and imagine a very low frequency supply. As a capacitor charges, a voltage builds up on its plates, which opposes the supply voltage. This is why the charge current of the capacitor is at first very high and then tapers off to zero as the capacitor becomes charged.

If the frequency of the AC supply is high enough (polarity reversal is fast enough) then the capacitor will be in the early part of its charge cycle, where the current is greatest when a polarity reversal takes place.

A high frequency will then cause the capacitor to charge and discharge in the early part of its first time constant period and this will cause a greater current to flow in the circuit for the same supply voltage.

We have deduced that capacitive reactance (X_C) is dependent on the frequency of the AC supply. In fact, we have deduced that the higher the frequency of the AC supply the more current flows and therefore the lower the capacitive reactance.

A larger capacitance would allow higher charge currents to flow. Therefore, changing the capacitance to one of a higher value would increase the current and therefore decrease the capacitive reactance.

The two factors determining capacitive reactance are frequency and capacitance.

EQUATION FOR CAPACITIVE REACTANCE

$X_C = 1/2\pi fC$ Where:
X_C = capacitive reactance in ohms.
2π = a numeric constant.
f = frequency in hertz.
C = capacitance in farads.

Look at the equation. $2\pi fC$ is in the denominator. 2π is a constant (it does not change).

We can say then, that capacitive reactance is inversely proportional to both frequency and capacitance.

In other words, if either capacitance or frequency was to double then capacitive reactance would halve. If capacitance or frequency was to halve then, the capacitive reactance would double and so on.

EXAMPLE CALCULATION

A capacitor of 0.05μF is connected to a 10 volt AC supply that has a frequency of 500kHz. What is the capacitive reactance in ohms and how much current will flow in the circuit?

$X_C = 1/2\pi fC$

$X_C = 1/(6.2831 \times 500 \times 10^3 \times 0.05 \times 10^{-6})$
$X_C = 6.366$ ohms

The current is found from Ohm's law by substituting X_C for R in the equation.

$I = E/X_C = 10/6.366 = 1.57$ amps

If we are given the capacitive reactances of a circuit and you need to find the net total capacitive reactance, use the same rules as for finding the total resistance of a circuit. For example, two capacitive reactances of 100 ohms in series are equal to 200 ohms and in parallel, the same combination would be 50 ohms.

To reinforce what we have learned have a look at the circuit shown in figure 13-2.

We could deduce the voltage across the 1μF and 2μF capacitors using the proportion method. Instead, let's work out how much current is flowing in the circuit, then using Ohm's law work out the voltage across each of the capacitors.

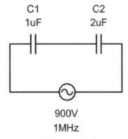

Figure 13-2

Reactance of the 1μF capacitor:

$X_{C1} = 1/2\pi fC$
$X_{C1} = 1/(6.2831 \times 1 \times 10^6 \times 1 \times 10^{-6})$
$X_{C1} = 0.159154943$ ohms

Reactance of the 2μF capacitor (it should be half as much because the capacitance is double):

$X_{C2} = 1/2\pi fC$

$X_{C2} = 1/(6.2831 \times 1 \times 10^6 \times 2 \times 10^{-6})$

$X_{C2} = 0.079577471$ ohms

The total capacitive reactance is the sum of the reactances:

X_C (total) $= X_{C1} + X_{C2} = 0.159154943 + 0.079577471$ ohms

X_C (total) $= 0.238732414$ ohms

The current flowing in the circuit is:

$I = E/X_C$ (total) $= 900/0.238732414 = 3769.911194$ amps

The voltage across the 1μF (it should be 600 volts):

$E = IX_{C1} = 3769.911194 \times 0.159154943 = 599.9998392$ volts

The voltage across the 2μF (it should be 300 volts):

$E = IX_{C2} = 3769.911194 \times 0.079577471 = 299.9999987$ volts

So C1 the smallest capacitor (1μF) has 600VAC RMS across it and C2 (2μF) has 300VAC RMS across it.

What if the circuit were DC? What would be the voltage across C1 and C2 in figure 13-3?

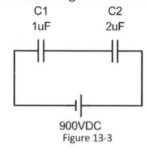

Figure 13-3

Well, we no longer have a frequency as it is DC; however, we do know that the highest voltage will be across the smallest capacitor and the lower voltage across the larger and that the sum of these voltages must equal 900V. Why?

Figure 13-3 is a series circuit. The current is the same in all parts of a series circuit. This means the amount of charge (Q) on C1 and C2 must be the same. For the smaller capacitor C1 to have the same charge on it as the larger C2, there must be more voltage across C1. (double the voltage in this case).

Q=CE and Q is the same for both capacitors. We also know the C1 is half the value of C2.
For C1 to have the same charge as C2, C1 must have twice the voltage of C2. Logically then; C1=600V and C2=300V and the sum, of course, the applied voltage of 900V.

As you can see from the graph in figure 13-4, as frequency increases capacitive reactance decreases and as capacitance increases capacitive reactance decreases.

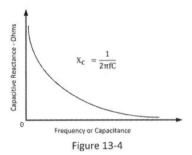

Figure 13-4

REACTANCE VERSUS F AND C

Capacitive reactance (X_C) is inversely proportional to both frequency and capacitance. Now if you cannot see that, without graphs, you have forgotten to "read" equations that we did back in the early chapters.

$X_C = 1/2\pi fC$

This equation should be speaking to you. The $2\pi fC$ is in the denominator. The right-hand side of the equal sign in this equation is a fraction. It has a numerator (1) and a denominator $2\pi fC$. When the denominator becomes larger in any fraction the entire fraction (which is X_C) becomes smaller.

So, at a glance, even if you do not have a clue what the symbols mean you should recognise that X_C is inversely proportional to frequency and capacitance.

That capacitive reactance decreases with increasing frequency becomes important later. As an example for now many amplifiers; transistors, FETs or valves have unwanted internal capacitance. At low frequencies the reactance of this internal capacitance is so high that this is not a problem. However as we move to higher frequencies the lower capacitive reactance becomes a real problem and ultimately determines the highest operating frequency of the amplifier.

POWER IN A CAPACITIVE CIRCUIT

Capacitance does not dissipate power. The only circuit property that can dissipate power is resistance.

All of the power taken from a purely capacitive circuit during the charge cycle is returned during the discharge cycle. So a pure capacitance does not dissipate any power. In practice, every capacitor has some resistance in its leads and plates. Dielectric leakage and dielectric loss also dissipates some power.

In practical capacitors, a small amount of power is lost in the dielectric, called dielectric loss. The atoms within the dielectric of a capacitor are placed under stress and do move slightly due to the electric field. Particularly at higher frequencies, the dielectric loss can become significant.

The losses in a "real" capacitor are due to the small amount of resistance in the leads and the plates - this is called resistive losses. Dielectric loss accounts for the greatest amount of loss. All losses increase with increasing frequency. Dielectric loss is caused by the alternating stress placed on the dielectric's atomic structure. In industry, there are many processes that utilise dielectric heating. Food is a lossy dielectric and can

90

be cooked through dielectric heating. Plastics can be welded. Laminated wooden furniture is shaped in a press and cured through dielectric heating.

Note: "real" means - a real world or actual capacitor as opposed to a pure capacitor which only exists on an engineering drawing.

ENERGY STORED IN A CAPACITOR

When a capacitor is charged by a supply and then removed from that supply, the negative plate has an excess of electrons and the positive plate a deficiency of electrons. There exists between the plates an electric field and this field places the molecules in the dielectric under stress. If a charged capacitor is placed across a lamp of the correct voltage the lamp will give off light as electrons move from the negative plate to the positive plate. This continues until there is no imbalance of charge on the plates.

While discharging through the lamp it gives off heat and light energy. There is no doubt then that a capacitor can store energy and later release that energy. Charge is stored on the plates of the capacitor and energy is stored in the electric field.

The amount of energy stored in a capacitor can be calculated in joules (watt-seconds) using the formula: -

$$E_n = \frac{CE^2}{2}$$

where E_n = Energy in joules (watt-seconds)
C = Capacitance (farads)
E = Voltage.

Most circuits we do not use very large capacitances with the exception of power supplies. I did experiment with some 6 x 700F super-capacitors in series and charged them to 15 volts. I found the super-capacitors stored enough energy to start a car engine several times. With the battery removed of course.

QUANTITY OF CHARGE IN A CAPACITOR

The charge of a capacitor is a measure of the imbalance of electrons on the plates. The Coulomb is the unit of charge. One coulomb is equal to a charge of 6.25×10^{18} electrons.

The amount of charge in coulombs is given by:-

$$Q = CE$$

where Q = charge in coulombs
C = capacitance in farads
E = voltage in volts.

WORKING VOLTAGE

The working voltage of a capacitor is the maximum voltage at which a capacitor operate without excessive current leaking or arcing through the dielectric. The working voltage is usually rated as so many volts DC. A capacitor might be marked at 1µF 600 VDC. This capacitor can withstand up 600V of DC electrical pressure. In AC circuits the capacitor should be used at no more than half the DC working voltage. Unless of course the capacitor has an AC working voltage on it.

If we had 2 x 1μF 600V capacitors and we connected them in series then we would have the equivalent to 0.5μF and working voltage of 1200V. If we were to have a string of many capacitors in series then you might think we can just keep adding up the working voltage, and you can, but precautions are necessary to make sure the voltage is shared evenly with all the capacitors. See equalising resistors.

If we had 2 x 1μF 600V capacitors and we connected in parallel we have the equivalent of 2μF 600V - the working voltage does not change.

Suppose we charged a capacitor to a certain charge Q and removed the charging source. We will pretend that our capacitor is perfect in that it has no losses. With the charging source removed the charge remains on the plates. We also have a certain voltage on the plates.

Now if we could move the plates of the capacitor apart without discharging the capacitor what would happen to the voltage on the plates?

If the plates are moved apart then C is decreased. Q stays the same as the charge is trapped on the plates. What happens to E?

We can transpose Q=CE for E and we get E=Q/C.

If Q is constant and C is decreasing the E must be increasing!

What we have just observed in albeit a mind experiment is that a capacitor can amplify voltage! Capacitors are used at very high frequencies as voltage amplifiers using this method. The capacitor is no ordinary capacitor. We use a semiconductor capacitor called a Varactor Diode. These amplifiers are called parametric amplifiers - meaning "amplification by variation of a parameter". These amplifiers are used where there are extremely weak signals such as in a radio telescope. Parametric amplifiers are very low noise. This noise is often reduced further by super cooling the amplifier.

Summary of capacitance so far: -

Capacitance is the property of an electric circuit that opposes changes of voltage. The unit of capacitance is the Farad.
If a charge of 1 coulomb produces a potential difference of 1 volt across the plates of a capacitor, then the capacitance is 1 farad.
Current flows into and out of, but not through, a capacitor.
The capacitance of a capacitor is directly proportional to the area of the plates and the dielectric constant and inversely proportional to the distance between the plates.
The opposition to current flow in a capacitive circuit is called Capacitive Reactance and is measured in ohms.
Capacitive reactance is inversely proportional to both frequency and capacitance. A purely capacitive circuit does not dissipate any power.
Current leads the voltage in a capacitive circuit.
Capacitors in parallel are treated like resistors in series.
Capacitors in series are treated like resistors in parallel.
Capacitors are given a voltage rating, which if exceeded, could cause the dielectric to conduct, destroying the capacitor.

14 - Inductance

Coils of wire were mentioned in the chapter on magnetism when we discussed the magnetic field about a coil carrying a current. An equally important aspect of the operation of a coil is its property to oppose any change in current through it. This property is called inductance.

Inductance is that property of an electric circuit that opposes changes of current.

Not to be confused with capacitance, which is, the property of an electric circuit to oppose changes of voltage.

When a current of electrons starts to flow along any conductor, a magnetic field starts to expand from the centre of the conductor outward. These lines of force move outward, through the conducting material itself and then continue into the air. As the lines of force sweep outward through the conductor, they induce an emf in the conductor. This induced voltage is always opposing the current that produced it. Because of its opposing polarity, it is called a counter emf, or a back emf. This opposing affect is called Lenz's Law and it is much like one of Newton's law's; every action has an equal and opposite reaction.

The effect of this backward pressure built up in the conductor is to oppose the immediate establishment of maximum current. It must be understood that this is a temporary condition. When the current eventually reaches a steady value in the conductor, the lines of force will no longer be in the process of expanding or moving and a counter emf will cease to be produced. In other words, there will be no relative motion between the conductor and the magnetic field. At the instant when current begins to flow, the lines of force are expanding at their greatest rate and the greatest value of counter emf will be developed. At the starting instant, the counter emf value almost equals the applied source voltage.

Current is minimum at the start of current flow. As the lines of force move outward, the number of lines of force cutting the conductor per second becomes progressively smaller and the counter emf becomes progressively less. After a period of time the lines of force expand to their greatest extent, the counter emf ceases to be generated and the only emf in the circuit is that of the source. Maximum current can now flow in the wire or circuit since the inductance is no longer opposing the source voltage.

This property of a coil or more correctly an inductor to oppose changes of current by self-inducing an opposing (counter) emf is called inductance. The unit of inductance is the Henry and the symbol for inductance is 'L'.

SELF INDUCTION

When the switch in a current carrying circuit is suddenly opened, an action of considerable importance takes place. At the instant the switch breaks the circuit, the current due to the applied voltage would be expected to cease abruptly. With no current to support it, the magnetic field surrounding the wire will collapse back into the conductor at a fast rate, inducing a high amplitude emf in the conductor. Originally, when the field built outward, a counter emf was generated. Now, with the field collapsing inward, a voltage in the opposite direction is produced. This might be termed a counter-counter emf, but is usually known as a self-induced emf. This self-induced emf is in the direction of the applied source voltage. Therefore, as the applied voltage is disconnected, the voltage due to self-induction tries to establish current flow through the circuit in the same direction, aiding the source voltage. The inductance induces a voltage to try and prevent the circuit current from decreasing. With the switch open it might be assumed that there is no path for the current, but the induced emf immediately becomes great enough to ionise the air at the opened switch contacts and a spark of current appears between them. Arcing lasts as long as energy stored in the magnetic field exists. This energy is dissipated as heat in any circuit resistance and the arc radiates energy as electromagnetic waves.

With circuits involving low current and short wires, the energy stored in the magnetic field will not be great and the switching spark may be insignificant. With long lines and heavy currents, inductive arcs many centimetres long may form between opened switch contacts on some power lines. The heat developed by arcs tends to melt the switch contacts and is a source of difficulty in high voltage high current switching circuits.

In a previous chapter, I gave you an example of the capacitance of a telephone line. A telephone line also has inductance. The normal operating voltage of a telephone line is about 50 volts DC. When a telephone line is suddenly open circuited by a technician or a linesman, the self-induced voltage on the line can be in the order of 2000 volts and this will produce a harmless but significant electric shock. The shock is harmless, as the amount of current is extremely small.

Remember, regardless of how current changes in a circuit containing inductance, the induced emf created by the inductance will oppose the change of current. The self induced emf of an inductor is often referred as a 'counter' or 'back emf'.

The unit of measurement of inductance is the Henry, defined as the amount of inductance required to produce an average counter emf of one volt when an average current change of one ampere per second is under way in the circuit. Inductance is represented by the symbol L in electrical problems and henrys are indicated by the letter 'H'.

COILING A CONDUCTOR

It has been explained that a piece of wire has the ability to produce a counter emf and therefore has a value of inductance. A small length of wire will have an insignificant value of inductance by general electrical standards. One henry represents a relatively large inductance in many circuit applications, where milli, micro or nanohenries are typically more practical. A straight piece of No. 22 wire one meter long has about 1.66 µH. The same wire wound onto an iron nail, or other high permeability core may produce 50 or more times that inductance.

A given length of wire will have much greater inductance if wound into a coil. If two loops are wound on the same conductor but a distance apart then each loop will have a small inductance. If the two loops are wound together side by side, they will collectively have twice the inductance of one loop.

When ten loops are wound next to each other as shown in figure 14-1, with the same current flowing there will be ten times the number of magnetic lines of force cutting each turn. Compared to 1 turn, this coiled inductor would produce a counter emf, ten times greater.

94

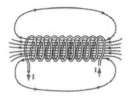

Figure 14-1

If the turns are stretched out, the field intensity will be less and the inductance will be less. Stretching the coil but keeping the same number of turns increases the length of the magnetic circuit and the inductance decreases.

The larger the radius or diameter of the coil, the longer the wire used and the greater the inductance. In single layer air core coils with a length approximately equal to the diameter, a formula that will give the approximate inductance in microhenries is:

$$L(\mu H) = \frac{r^2 N^2}{24r + 25l} \quad \text{Equation 14-1}$$

Where:

L = Inductance in μH.

N = number of turns.

r = radius in cm.

l = length in cm

The inductance of straight wires is found in antennas, in power lines and in ultra-high frequency equipment. In most electronic and radio applications where inductance is required, space is limited and the wire is wound into either single layer or multilayer coils with air, mixed powdered iron compound (ferrite), or laminated iron (many thin sheets) cores. The advantage of multilayer coil construction for high values of inductance becomes obvious when it is considered that, while 2 closely wound turns produce 4 times the inductance of 1 turn, the addition of 2 more turns closely wound on top of the first 2 will provide almost 16 times the inductance. The direction of winding has no effect on the inductance value of a given coil.

In many applications, coils are constructed with ferrite cylinders (slugs) that can be screwed into or out of the core space of the coil. This results in a controlled variation of inductance, maximum when the ferrite core "slug" is in the coil and minimum with it out.

A particular type of coil is the toroid. It consists of a doughnut shaped ferrite core, either single or multilayer wound as shown in figure 14-2. Its advantages are high values of inductance with little wire and therefore little resistance in the coil and the fact that all the lines of force are in the core and none outside (provided there is no break in the core). As a result, it requires no shielding to prevent its field from interfering with external circuits and to protect it from effects of fields from outside sources. Two toroids can be mounted so close that they nearly touch and there will be almost no interaction (magnetic coupling) between them. Inductors come in a huge number of packages from tonnes to a few grams in weight. With some packages, it is not evident that the device is an inductor as they are small and encapsulated.

Figure 14-2

There are so many different packages for inductors it is too much to show them all. Some look very much like resistors. Others are copper tracks etched into a circuit board. Others may be formed with a length of transmission line.

Figure 14-3

Schematic symbols of inductors.

Left to Right. Fixed-air core, tapped-air core, variable, iron core, ferrite core.

THE TIME CONSTANT OF AN INDUCTANCE

The time required for the current to rise to its maximum value in an inductive circuit after the voltage has been applied will depend on both the inductance and the resistance in the circuit. The greater the inductance, the greater the counter emf produced and the longer the time required for the current to rise to maximum. With a constant value of inductance in a circuit and more resistance, then less current can flow.

As with capacitive and resistive circuits, the time required for the current to rise to about 63% (more precise 63.2%) of the maximum value (called the time constant) can be determined by:

$T = L/R$

Where
T = time in seconds.
L = inductance in henrys (H).
R = resistance in ohms.

EXAMPLE

Calculate the time constant of a 10 henry inductance with 10 ohms of resistance.

$T = L/R = 10/10 = 1$ second

Suppose the emf applied to this inductor was 10 volts. What would be the current flowing after 1 second?

The final current after 5 time constants will be:

$I = E/R = 10/10 = 1A$

After 1 time constant the circuit current will have reached 63.2% of what will be its final value, or 63.2% of 1A = 632mA.

CURRENT LAGS VOLTAGE IN AN INDUCTIVE CIRCUIT

In a resistive circuit, the current and voltage obey Ohm's law. If you increase the voltage across a resistance, the current increases straight away to the new Ohm's law value determined by I=E/R. In an inductive circuit, it takes time for the current to build up to its final value. The current lags behind the voltage. The current will eventually reach its Ohm's law value after 5 time constants. However, if the voltage is continually changing as in an AC circuit, the current will always lag the voltage across the inductor by 90 degrees.

Think of 'L' for inductance and 'L' for lag - current lags voltage in an inductive circuit.

THE ENERGY IN A MAGNETIC FIELD

The current flowing in a wire or coil produces a magnetic field around itself. If the current suddenly stops, the magnetic field held out in space by the current will collapse back into the wire or coil. Unless the moving field has induced a voltage and current into some external load circuit, all the energy taken to build up the magnetic field will be returned to the circuit in the form of electric energy as the field collapses.

The amount of energy stored in the magnetic field of an inductor is given by;

$$E_n = \frac{LI^2}{2}$$

Where

E_n = energy stored in joules (watt-seconds)
L = inductance in henries
I = current in amperes

CHOKE COILS

The ability of a coil to oppose any change of current can be used to smooth out varying or pulsating types of current. In this application, an inductor is known as a choke coil since it chokes out variations of amplitude of the current. For radio frequency (RF) AC or varying DC, an air core coil may be used. For lower frequency circuits greater inductance is required. For this reason, iron core choke coils are found in audio and power frequency applications.

A choke coil will hold a nearly constant inductance value until the core material becomes saturated. When enough current is flowing through the coil to saturate the core magnetically, variations of current above this value can produce no appreciable counter emf and the coil no longer acts as a high value of inductance to these variations. To prevent the core from becoming magnetically saturated, a small air gap may be left in the iron core. The air gap introduces so much reluctance in the magnetic circuit that it becomes difficult to make the core carry the number of lines of force necessary to produce saturation. The gap also decreases the inductance of the coil. An air coil cannot be saturated.

DEFINITION OF A HENRY

If a coil has a rate of change of current of 1 ampere per second and this produces a counter emf of 1 volt, then the coil is said to have an inductance of 1 henry.

MUTAL INDUCTANCE

If one coil is placed near another so that the magnetic fields interact with each other, then the moving magnetic field in one will induce a voltage into the other. This ability of one coil to effect another is called mutual inductance.

The further apart the two coils are, the fewer the number of lines of force that interlink the two coils and the lower the voltage induced in the second coil.

The mutual inductance can be increased by moving the two coils closer together or by increasing the number of turns of either coil.

When all the lines of force from one inductor are linked to another, unity coupling is said to exist and the mutual inductance is:

$$M = \sqrt{L1 \times L2}$$

Where M is the mutual inductance in henries.

The above formula assumes 100% coupling between the two inductors L1 and L2. This equation assumes that all the magnetic lines of force from L1 cut all the turns of L2. If this is not the case then M is determined by:

$$M = k\sqrt{L1 \times L2}$$

Where k is the percentage of coupling.

For example: suppose a 5 henry coil has 75% of its lines of force cutting a 3 henry coil. What is the mutual inductance?

$$M = 0.75\sqrt{5 \times 3}$$
$$M = 2.9H$$

COEFFICIENT OF COUPLING

The degree of closeness of coupling of two coils can also be expressed as a number between 0 and 1 rather than as a percent. A percentage of 100 is equal to a coefficient of coupling of 1 or unity. 75% coupling is 0.75 as above. No coupling is 0.

The formula for mutual inductance can be transposed for coefficient of coupling:

$$k = \frac{M}{\sqrt{L1 \times L2}}$$

INDUCTANCES IN SERIES

When you add inductances in series, you are in effect simply increasing the number of turns on the inductor. Therefore, to find the total inductance, sum the individual inductances.

Lt= L 1 + L2 + L3 + ... n

INDUCTANCES IN PARALLEL

If the inductances are connected in parallel, the total inductance is calculated by using a formula similar to the parallel resistance formula.

1/Lt = 1/L 1 + 1/L2 + 1/L3 + ... n

Both of these equations assume that the magnetic lines of force from all the inductors are not coupled (linked) to the others i.e. the mutual inductance is zero. Another way of describing how inductances are linked is called coefficient of coupling (k). If k=O then there is no coupling. If k=L, the two inductances are completely coupled.

INDUCTIVE REACTANCE

It has been explained that DC flowing through an inductance produces no counter emf to oppose the current. With varying DC, as the current increases, the counter emf opposes the increase. As the current decreases, the counter emf opposes the decrease. Alternating current is in a constant state of change and the effect of the magnetic fields is a continually induced voltage opposition to the current. This reacting ability of the inductance to oppose a changing current is called inductive reactance. Inductive reactance is the opposition to current flow presented by an inductance in a circuit with changing current and is measured in ohms.

$$X_L = 2\pi f L$$

Inductive reactance is directly proportional to frequency and inductance.

Just as with resistance and capacitive reactance, the total inductive reactance in ohms is found using the same form of the equation.

In series:

X_L(total) = X_{L1} + X_{L2} + X_{L3} + ... n
and parallel:
$1/X_{L(total)}$ = $1/X_{L1}$ + $1/X_{L2}$ + $1/X_{L3}$ + ... n

The voltage across and the current through an inductive reactance can be determined using Ohm's law.

EXAMPLE

An inductance of 100µH is in series with an inductance of 200µH and connected to a 10 volt AC supply with a frequency of 500kHz. How much current will flow in this circuit?

We need to find the total inductive reactance of the circuit. We could find the individual reactances of each inductor and add them, or we could find the total inductance and then the total inductive reactance. I will do it using the latter method.

LT = L1 + L2 = 100µH + 200µH = 300µH

X_L= $2\pi f L$
X_L = 2π X 500 x 10^3 x 300 x 10^{-6}
X_L= 942Ω

I=E/R = 10/942 (x 1000 for milliamps) = 10.6mA

How much power is dissipated in the above example?
There is no resistance in the circuit, so no power is dissipated. You cannot/must not substitute X_L or X_C for R in the power equations. For example, P=I^2R is okay but P=$I^2 X_L$ is not. Reactance, either inductive or capacitive does not dissipate power - only resistance does.

Take a pure inductive reactance of 240 ohms and imagine it being connected to a power point. The current that flows would be I=E/ X_L = 240/240 = 1 ampere. Even though 1 ampere would flow in the circuit, no power

is dissipated as there is no resistance. All of the power supplied charging the inductor during one cycle is returned to the power point (the power supply) on the alternate half cycle. A pure inductor has no losses.
Only resistance dissipates power

Power authorities dislike reactive loads like this as current drawn does not register on the energy meter. The power lines to your house carry the charge and discharge 'reactive' currents do have resistance - the lines do dissipate power - and this power loss is not metered.

LOSSES IN A REAL INDUCTOR

A real world inductor does have losses. These losses include resistive losses of the windings and core losses. Hysteresis loss is a core loss caused by the continuous realignment of the magnetic domains. Eddy currents are small circulatory currents (eddies) induced into a conductive core. Core losses increase substantially with increasing frequency. Hysteresis and eddy current loss will be discussed in more detail in the chapter on transformers.

PHASE AND INDUCTANCE

In a purely inductive circuit (no resistance) the current lags the voltage by 90°. If there is inductance and resistance in the circuit then the phase angle between voltage and current will be less than 90° lagging. The actual phase angle will depend on the ratio of the inductive reactance and the resistance X_L/R.

For example suppose we had and inductive reactance of 30Ω and a resistance of 30Ω then the phase angle, that is, how much the current lags the voltage is given by phase angle $\theta = \tan X_L/R$ or $\tan(30/30) = 45°$ lagging.

In an LR circuit the phase angle can be between 0° to 90° lagging.

In a CR circuit the phase angle (from $\tan X_C/R$) can be between 0 to 90° leading. Leading means the current leads the voltage.

This lead and lag occurs because Inductance is the electrical property that opposes changes of current and capacitance is the electrical property that opposes changes of voltage. We will be doing more on phase angle later when we study LCR circuits.

The man that did the most work on Inductance was Joseph Henry and the unit of inductance is named after him.
Joseph Henry was the first to coil insulated wire tightly around an iron core in order to make a more powerful electromagnet. Henry shared his knowledge freely and to some extent has not been given full credit. Samuel Morse acquired much of the knowledge to build the Telegraph from Henry. It was Henry's invention of the Relay that made the Telegraph possible. Henry was appointed the first Secretary of the Smithsonian Institution in 1846 and served in this capacity until 1878. He did early pioneering work into the nature of sunspots. From 1832 to 1846, Henry served as the first Chair of Natural History at the

College of New Jersey (now Princeton University). While in Princeton, he taught a wide range of courses including natural history, chemistry, and architecture, and ran a laboratory on campus. Decades later, Henry wrote that he made several thousand original investigations on electricity, magnetism, and electro-magnetism while at the Princeton faculty.

15 - Meters

We do need to know about using moving coil meters to build ammeters, voltmeters and ohmmeters.

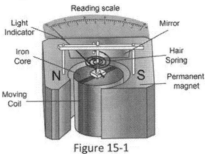

Figure 15-1

A pictorial diagram of the moving coil meter is shown in figure 15-1. The moving coil is wound on an aluminium former. The coil is free to move clockwise and anticlockwise through about 130 degrees, this motion in turn moves the pointer on the scale from left to right. Without any current flowing through the moving coil, the pointer will be towards the left of the scale, pushed there by a small coil return spring. The pointer will not go past the left end of the scale as there is a small pointer stop. When current is passed through the moving coil, a small magnetic field is created around it which interacts with the field created by the permanent magnet in such a way as to cause the moving coil and the pointer along with it to rotate clockwise. So we have a simple but very effective method of detecting current flow with this instrument. Importantly, the degree of deflection from left to right by the pointer is directly proportional to the current. If a certain current causes the pointer to move to 1/4 scale, then twice that current will cause the pointer to move to 1/2 scale - this is a direct proportion.

The other important feature of the moving coil meter is that it is extremely sensitive to very small currents, typically in the order of microamperes and therefore does not consume much power from the circuit in which it is placed. The moving coil meter is a microammeter. Typical current to enable full-scale deflection ranges from 1μA to 30mA. The moving coil does have some resistance of course. A movement with a smaller full-scale deflection current has a higher coil resistance, as more turns of fine wire are needed to obtain that deflection.

We can make use of this meter by calibrating and using it to measure current, voltage and resistance. From now on I will draw the moving coil meter as a circle, but it does help to remember what is inside.

The other point I would like to make is that many modern meters are digital. They do not use a moving coil. The digital circuitry performs an analogue to digital conversion of the applied current or voltage and displays the amount as a numerical readout. These meters are very accurate. However, the moving coil meter (called an analogue meter) has advantages that will see it in use for a long time to come. The greatest advantage of the analogue meter is its ability to show changes in voltage or current. For example, if an analogue meter is placed across a charging capacitor, the pointer will slowly rise up the scale as the capacitor charges. I have an all bells and whistles digital meter. When I place it across a charging capacitor, because the voltage is changing, the digital display just "blinks" as it is unable to show a varying voltage adequately.

MEASUREMENT OF CURRENT

Whether we are measuring amperes, milliamperes or microamperes, two important facts to remember are:

1. The current meter (ammeter) must be in series with the circuit in which the current is to be measured. The amount of deflection depends on the current through the meter. In a series circuit, the current is the same in all parts of the circuit. Therefore, the current to be measured must be made to flow through the meter as a series component in the circuit.

2. A DC meter must be connected with the correct polarity for the meter to read upscale. Reversed polarity would make the meter read down-scale, forcing the meter pointer hard against the left-hand stop.

An ammeter should have a very low resistance when compared to the circuit in which it is placed. An arbitrary figure is 1/100th of the resistance of the circuit in which the ammeter is placed.

AMMETER SHUNTS

A meter shunt is a precision resistor connected across the meter movement for the purposes of shunting or bypassing, a specific fraction of the circuit current around the meter movement. Shunts are usually inside the meter case. The schematic symbol for an ammeter does not show the shunt.

While shunts are called resistors, they are extremely low-value resistors. A high current shunt may be a brass bar and measure in the order of milli-ohms or less. Figure 15-2 shows a typical high current range shunt.

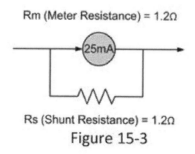

Figure 15-2

One way of adjusting precisely the resistance of a high current shunt is to make small cuts in it using a very fine metal saw.

For example, we may have a 25mA meter movement with a moving coil resistance of 1.2 ohms. We want the meter to be able to read 50mA full scale. In other words, to double the range of the meter from 0-25mA to 0-50mA.

To achieve this, a shunt resistance of 1.2 ohms (equal to the resistance of the moving coil) would suffice.

Rm (Meter Resistance) = 1.2Ω

Rs (Shunt Resistance) = 1.2Ω

Figure 15-3

With a 1.2-ohm shunt, half of the total current will flow through the meter and half through the shunt since both form a parallel circuit consisting of two 1.2 ohms resistances. The use of the shunt in this instance has extended the full-scale deflection (FSD) of the meter from 0-25mA to 0-50mA.

With a correct shunt, a moving coil meter can be used to measure any amount of current (FSD). The shunt is a low value resistor. In typical ammeters, the shunt is often a solid copper bar with a resistance in milli-ohms, particularly for high current ranges on the meter.

It is common to hear a current meter called an 'ampmeter' - this is wrong; the correct name for a current meter is an ammeter with no 'p'.

CALCULATING ANOTHER SHUNT

Figure 15-4 shows an ammeter. The moving coil has a resistance of 2000 ohms and is deflected full scale with 50 microamperes. We wish to modify the meter to measure 1 ampere full-scale deflection. Calculate the value of the shunt resistor that would extend the range of the ammeter to 0-1 amperes.

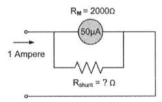

Figure 15-4

This is just an Ohm's law problem. Assume that one ampere is flowing in the circuit. You know there is 50uA flowing through the meter movement. The current through the shunt is then:

Ishunt = 1 amp - 50µA = 0.99995 amperes.

We now know the current through the shunt. What is the voltage across it? We know the resistance of the meter and the current through it. Since the meter and the shunt are in parallel, they will have the same voltage. So, if we calculate the voltage across the meter we will know the voltage across the shunt:

Eshunt = Im x Rm= 50µA x 2000 ohms= 0.1 volts.

Now calculate the resistance of the shunt:

Rshunt = Eshunt / Ishunt = 0.1 / 0.99995 = 0.100005 ohms.

This shunt, which is just over 1/10th of an ohm, would be a solid bar inside the meter. The manufacturer (or the builder) would use a reference current and adjust the resistance of the shunt accurately.

VOLTMETER

Although a meter movement responds only to current moving in the coil, it is commonly used for measuring voltage by the addition of a high resistance in series with the movement. Such a high series resistance is called a *multiplier*.

The multiplier must be higher than the coil resistance in order to limit the current flow through the coil. The combination of a meter movement with its added multiplier then forms a voltmeter.

Since a voltmeter has a high resistance, it must be connected in parallel to measure the potential difference between any two points in a circuit. Otherwise, the high

resistance multiplier would add so much series resistance to the circuit the current would be reduced to a very low value.

The circuit is not opened when you connect a voltmeter in parallel. Because of this convenience (not having to break the circuit), it is common to make voltmeter tests when troubleshooting rather than ammeter tests. If you need to know the current through a resistor, it is far easier to measure the voltage across it and then work out the current using Ohm's law.

Figure 15-5 shows how a 1mA meter movement is connected with a multiplier to enable it to be used as a voltmeter with an FSD of 10 volts. In other words, a voltmeter with a 10V range.

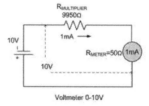

Figure 15-5

With the 10V applied by the battery, there must be a total of 10,000 ohms of resistance to limit the current to 1mA for full-scale deflection (FSD) of the meter movement. Since the movement has a resistance of 50 ohms, 9950 is added in series, resulting in a total resistance of 10kΩ. Current through the meter is then: I=10/10,000 =1mA.

With 1mA in the movement, the FSD can be calibrated as 10V on the meter scale. Of course, since deflection is directly proportional to current the meter can be calibrated proportionately from 0-10 volts. For example, half scale would be 5 volts, 1/4 scale 2.5 volts, etc.

With the battery removed the circuit now consists of the meter movement and the multiplier. This is our 0-10V voltmeter. Different multipliers can be switched in for other voltage ranges.

Let's calculate another multiplier to make sure we understand this.

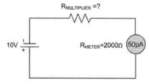

Figure 15-6

The circuit of figure 15-6 is a voltmeter. It has a very sensitive meter movement (50µA). What is the value of the multiplier resistor required to allow this meter to measure from 0 to 10 volts?

We know that at full-scale deflection (FSD) the current in the circuit and through the multiplier will be 50μA. We can work out the voltage across the meter movement and subtract this from the applied voltage (10V) to get the voltage across the multiplier.

voltage across meter= I meter x Rmeter = 50uA x 2000 ohms= 0.1 volts

Rmultiplier = Emultiplier / I multiplier

Rmultiplier = (10-0.1 volts)/ 50μA = 9.9 volts/ 50uA = 198,000Ω

VOLTMETER RESISTANCE

The high resistance of a voltmeter with a multiplier is essentially the value of the multiplier resistance. Since the multiplier is changed for each range, the voltmeter resistance changes. Most moving coil voltmeters are rated in ohms of resistance needed for 1V of deflection. This value is the Ohms-Per-Volt rating of the voltmeter.

Figure 15-7 is a circuit of a multi-range voltmeter. Why don't you check this circuit yourself by calculating two or three of the multipliers?

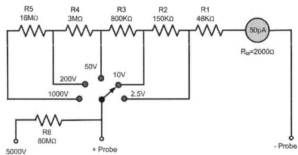

Figure 15-7

The leads on the voltmeter are shown by the+ and - signs. When this voltmeter is switched to the 2.5 volt scale, can you see that the only multiplier resistance used is the 48kΩ. So the total resistance of the voltmeter is 48kΩ + 2000 ohms =50kΩ.

The total resistance of this voltmeter on the 2.5 volt scale is 50kΩ.

If we divide the total resistance of the voltmeter by the scale to which it is switched, we get the sensitivity of the voltmeter in ohms-per-volt:

ohms-per-volt=total resistance/ scale= 50kΩ / 2.5 = 20,000 ohms-per-volt.

This sensitivity works for all other scales as well. If we want to know the total resistance of the voltmeter on any scale, just multiply the sensitivity by that scale.

Example: What is the total resistance of this voltmeter if it is switched to the 10 volt scale?

Total resistance of voltmeter= ohms-per-volt x scale.
Total resistance of voltmeter= 20,000 x 10 = 200,000 ohms.

Is this right?

Check it in the circuit. On the 10 volts scale the voltmeter's resistance consists of: 150kΩ+ 48kΩ+ 2000Ω= 200,000 ohms.

Using this method, you can work out quickly the total resistance of any voltmeter on any scale - provided you know the ohms-per-volt.

A QUICK WAY TO WORK OUT SENSITIVITY

A quick way of working out the ohms-per-volt or sensitivity of any voltmeter is to take the reciprocal of the full-scale deflection current of the movement. In the last example, current FSD is 50uA and so the reciprocal of 50uA is $1/50 \times 10^{-6}$ which gives 20,000 ohms-per-volt.

We need to know the sensitivity of a voltmeter so that we can determine if it is going to affect significantly the circuit in which it is placed.

Suppose we were using a 20,000 ohms-per-volt voltmeter on a 10 volt range. The voltmeter's resistance would be 20,000 x 10 = 200kΩ. So our voltmeter has a resistance of 200,000 ohms when switched to the 10 volt range. When we place our voltmeter in parallel with a component to measure the voltage across that component, the voltmeter is adding 200,000 ohms of resistance across the circuit.

Now, if the component we are measuring the voltage across is a 1000Ω resistor, placing the voltmeter resistance of 200kΩ in parallel with 1000Ω is not going to disturb the resistance of the circuit much at all. However, if we were to use the same voltmeter across a 200kΩ resistor, we would be reducing the resistance of the 200kΩ resistor to 100kΩ. 200kΩ in parallel with 200kΩ is 100kΩ.

This does not always matter as long as you know what effect the voltmeter is having

on the circuit. In some cases, though, the interaction of the voltmeter with the circuit may cause the circuit to stop operating, particularly in high resistance radio frequency circuits.

Very good voltmeters have a very high resistance in the vicinity of 10MΩ-per-volt. Whether or not you have to be concerned about the interaction of the voltmeter and the circuit still depends on the circuit from which you are taking the measurement.

Most digital meters have an input resistance which is not scale dependent. So you may see a digital voltmeter with 10MΩ resistance irrespective of the voltage scale used.

OHMMETERS

An ohmmeter consists of an internal battery, the meter movement and a current limiting resistance. When measuring resistance, the ohmmeter leads are connected across the resistance to be measured. The circuit power is turned off. Only with the power off can you be sure that it is only the ohmmeter's battery that is producing current and deflecting the meter movement. Since the amount of current through the meter depends on the external resistance, the scale can be calibrated in ohms.

The amount of deflection on the 'ohms' scale indicates the measured resistance directly. The ohmmeter reads up-scale regardless of the polarity of the leads because the polarity of the internal battery determines the direction of current through the meter movement.

The leads of a meter are normally coloured 'black' for negative and 'red' for positive. It is important to remember that a multi-function meter, when switched to ohms, may have the output polarity reversed, because of the internal battery or cell, supply a positive voltage to the negative lead and vice-versa for the other lead. This can be of particular importance when testing semiconductors.

The ohmmeter circuit shown in figure 15-8 has 1500 ohms of resistance. The 1.5 volt cell will then produce 1mA if the leads are shorted and the pointer will go to full scale. In a practical ohmmeter, a small variable resistor would be in series to 'zero' the meter exactly.

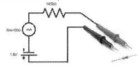

Figure 15-8

When the ohmmeter leads are short-circuited the meter will show full scale or zero resistance. With the ohmmeter leads open (not touching), the current is zero and the ohmmeter indicates infinite resistance.

Therefore, the meter face can be marked zero ohms at the right for full-scale deflection and infinite ohms on the left for no deflection. The in between values can be marked in ohms by calibrating the meter against known resistances.

Notice with an ohmmeter the red lead is negative and the black positive. When this meter is switched to volts or amps, the lead colours are correct as voltage or current is coming from outside the meter and going in. With an ohmmeter, the current and voltage is coming out of the meter. This is handy to know when testing semiconductors in the circuit as the polarity of the leads can be used to reverse or forward bias PN-junctions.

MULTIMETERS

All of the above meters can be built into the one box. Shunts, multipliers and the battery can be switched in and out of the circuit as required to perform the various functions. Multimeters can be digital or analogue. There are advantages and disadvantages to each. Digital meters make static (non-changing) readings clear. Analogue meters make changing voltages and currents easy to read. Any good digital meter will have auto ranging and over volt and current protection. For an investment of a few dollars, you can buy a cheap multimeter and just do some measurements around the workshop or the house - you will learn a lot.

A word of caution, though; cheap meters like the one shown in 15-9 are great to learn on and you really do not care if you destroy if for the price - but - be very careful using such meters on high voltages. They tend to have cheap probes and a brittle plastic case. As long as you are careful, you can learn a lot from the meter and when you graduate to a better one it can be used on your car electrics.

Figure 15-9

Great for learning with but be careful of high voltages; these cheap meters are not safe on mains electricity regardless of what the manufacturer says.

When an analogue moving coil meter is not in use it is good practice to short circuit the terminals and put the meter on a low current scale. This has the affect of physically dampening the meter from being jostled around. The meter is a coil on a pivot in a magnetic field, in effect a simple motor. If the meter is bumped the needle with the coil will attempt to deflect. However, in doing so there will be a counter emf induced into the coil that opposes this motion. A backward torque is electrically produced that causes the meter movement to be damped.

Moving coil meters often have a mirrored scale. When reading the meter look over the top of it straight down so that you line up the pointer with its reflection in the mirrored scale. This helps you to avoid parallax error.

DIGITAL MULTIMETER (DMM)

Digital Multimeters range in price from a few tens of dollars to hundreds of dollars. Modern meters have many functions other than volts, amps and ohms. Some can measure capacitance, frequency and temperature. There are auto ranging meters. You just have to switch to volts or whatever you want and the meter will look after the range.

Good meters have a large backlit display and update the display very quickly. Most DMMs have a bargraph in the digital readout to simulate an analogue display. Nearly all have a continuity test that makes an audible alarm when the circuit is complete.

A meter marked TRUE RMS will read the true RMS value of a voltage or current irrespective of wave shape being measured.

Electrical measuring instruments are assigned a CAT number. For example CATI, CATII and CATIII. These categories tell you about the safe working voltage and insulation of the meter. Caution needs to be exercised as some imported meters can be falsely labelled.

The DMM shown in figure 15-10 is CATIII Measurement Category III:

This category refers to measurements on hard-wired equipment in fixed installations, distribution boards, and circuit breakers. Other examples are wiring, including cables, bus bars, junction boxes, switches, socket outlets in the fixed installation, and stationary motors with permanent connections to fixed installations.

Safety when working on high voltages is paramount. A CATIII meter should be used with the correct leads and probes. (figure 15-10)

16 - Electric Cells & Batteries

There are many different types of batteries. The basic principles of operation are all the same. There is much material here that you do not need to know for examination. It might seem a bit strange, but let's start off our discussion about batteries by starting with corrosion. The two topics are related chemically and knowledge of corrosion is useful particularly when it comes to antennas.

CORROSION

Corrosion is a chemical reaction. Corrosion involves removal of metallic electrons from metals and the formation of more stable compounds such as iron oxide (rust), in which the free electrons are usually less numerous. In nature, only rather chemically inactive metals such as gold and platinum are found in pure or nearly pure form; most others are mined as ores that must be refined to obtain the metal. Corrosion simply reverses the refining process, returning the metal to its natural state. Corrosion compounds form on the surface of a solid material. If the compounds are hard and impenetrable and if they adhere well to the parent material, the progress of corrosion is arrested. If the compound is loose and porous, however, corrosion may proceed swiftly and continuously.

Aluminium is a good example. When aluminium corrodes, it becomes covered with aluminium oxide. Aluminium oxide adheres well to aluminium and prevents further corrosion. This makes aluminium an excellent choice for antennas. An added bonus is its light weight and strength.

HOW THINGS CORRODE

If two different metals are placed together in a solution (electrolyte), one metal will give up ions to the solution more readily than the other. This difference in behaviour will bring about a difference in electrical voltage between the two metals. If the metals are in electrical contact with each other, electrons will flow between them and they will corrode; this is the principle of the galvanic cell or battery. Though useful in a battery, this reaction causes problems in a structure. For example, steel bolts in an aluminium framework may, in the presence of rain or fog, form multiple galvanic cells at the point of contact between the two metals, corroding one or both them.

An electrolyte is any liquid that conducts electricity. With corrosion, the electric cell effect of dissimilar metals in the presence of an electrolyte is undesirable. Nevertheless this is what we desire in an electric cell. Although the term battery, in strict usage, designates an assembly of two or more voltaic cells capable of such energy conversion, it is commonly applied to a single cell of this kind.

THE VOLTAIC CELL

When two different electrodes are immersed in an electrolyte, the chemical reaction which takes place results in a separation of charges. The arrangement required to convert chemical energy into electrical energy is called the voltaic cell.

The charged conductors are the electrodes, serving as the connection of the cell to an external circuit. The potential difference resulting from the separated charges enables the cell to function as the source of applied voltage.

Electrons from the negative terminal of the cell flow through the external circuit and return to the positive terminal. The chemical action in the cell continuously separates charges to maintain the terminal voltage that produces the current in the circuit.

AN EXPERIMENT

If you have a voltmeter with a low DC volts scale, you can easily demonstrate the action of a Voltaic cell. All you need are two different conductors and some type of electrolyte. I have had great success using a lemon or an orange. Just take the lemon and push two electrodes into it. I have used different nails, paper clips; almost anything will work. Place your voltmeter across the conductors and you will measure a potential difference, usually a significant fraction of a volt. You could find out if two coins are made of the same alloys or not! The amount of voltage developed by two coins you can tell (roughly) how dissimilar they are. No doubt this technique has an application in metallurgy.

SEPARATION OF CHARGES

When metals dissolve in an electrolyte, the chemical action causes separation or disassociation of the molecules, which results in charged ions. Figure 16-1 shows a generic cell with two electrodes immersed in an electrolyte which could be liquid or paste. A chemical reaction with the electrodes causes a flow of ions through the electrolyte. This flow of ions to the two electrodes causes one to become negatively charged and the other to be positively charged.

As long as the reaction continues and the external circuit can be connected between the negative and positive terminals an electron flow will occur between them and do work for us. In this case, the electron flow is lighting (by current causing heat) a light bulb. The actual chemistry is not important to us as cells use many different types of electrodes and electrolytes. It is the principle of the disassociation of charge due to ionic movement in the electrolyte that makes one electrode negative and the other positive.

The ionic current flow in the electrolyte is an "electric current". Electric current is not "just" the flow of electrons. Electric current is the ordered flow of any charge. One of the first practical batteries developed used zinc and copper electrodes immersed in ammonium chloride. These early cells were assembled using glass jars. To obtain the required output voltage, the cells (jars), were connected in series. These banks of cells connected together were called a Voltaic pile; we call it a battery today.

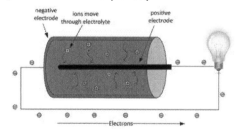

Figure 16-1

PRIMARY CELLS

In a primary cell, the chemical action of forming the solution is not reversible. For instance, zinc can dissolve in ammonium chloride, but the process cannot be reversed to form the zinc electrode from the solution.

SECONDARY CELLS

In secondary cells, the chemical action occurring in the electrolyte can be reversed. The electrodes can dissolve in the solution when the cell provides current to the external circuit. In this case, the cell is discharging. When an external voltage is applied to the cell to make current flow in the reverse direction, the metal comes out of the solution and is deposited back onto the electrodes, recharging the cell. Since a secondary cell can be recharged, it is also called a storage cell.

The carbon-zinc dry cell is a modern day version of the Leclanche cell. See figure 16-2. We will look at this cell with a view to understanding how most cells work in general. You do need to know a couple of things specific to other cells and these will be covered in this chapter

The Leclanche cell is a zinc-carbon cell. The carbon rod in the centre is the positive electrode and the zinc case is the negative electrode but also the housing for the entire cell. The electrolyte in the cell is ammonium chloride. The cell is a primary cell in that it cannot be recharged. The Eveready-Red is a common zinc-carbon cell.

As the cell discharges, the zinc electrode becomes dissolved in the ammonium chloride

electrolyte. The electrolyte is in the form of a paste. The cell is completely discharged when the electrolyte is converted to zinc ammonium chloride, which usually corresponds with the zinc, in the zinc case almost being depleted. Early carbon-zinc cells would often leak electrolyte into equipment when they were discharged, as the zinc case would get holes in it.

Construction of Zinc-Carbon cell (Leclanche)

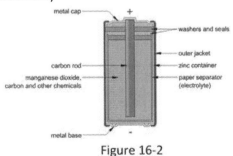

Figure 16-2

POLARISATION

Now, if you're old enough, you might remember when the zinc-carbon battery would no longer work. It could be heated by placing it in or near an oven for a few minutes and it would come to life again, for a while at least.

Well, one of the problems with the zinc-carbon cell is that as the chemical reaction takes place to charge both of the electrodes, hydrogen gas forms around the carbon rod in the form of small bubbles. This build up of hydrogen gas around the carbon rod will begin to insulate the carbon rod from the electrolyte. Since the surface area of the carbon rod in contact with the electrolyte is reduced, the terminal voltage of the cell will drop dramatically and the cell(s) will no longer operate the equipment. This unwanted build up of hydrogen gas (and other gases in other cells) is called polarisation.

Placing the dry cell in the oven would drive the hydrogen gas from the carbon rod bringing it back to life for a while. The higher temperature would also increase the rate of chemical reaction between the electrodes and the electrolyte.

A chemical called manganese dioxide is mixed in with the electrolyte. Manganese dioxide absorbs the hydrogen gas and prevents the gas from building up on the carbon rod. Manganese dioxide is rich in oxygen and this oxygen combines with the hydrogen to form water. Manganese dioxide in the zinc-carbon cell is called the depolariser, as removing hydrogen from the carbon rod is called depolarisation.

LOCAL ACTION

If the zinc electrode (the case) contains impurities, small voltaic cells are formed

which do not contribute to the output voltage of the cell. Also, these unwanted cells caused by impurities in the zinc, consume chemicals and dissolve the zinc. This is called local action. To minimise local action, the zinc electrode is coated with mercury by a process called amalgamation.

ALKALINE CELL

The carbon-zinc cell is now largely being superseded by the alkaline cell, which has better discharge characteristics and will retain more capacity at low temperatures. The alkaline cell typically has an ampere-hour capacity of about twice that of the carbon zinc cell. Its nominal voltage is also 1.5V.

The negative electrode is manganese dioxide and the positive electrode is zinc. The electrolyte is potassium hydroxide or sodium hydroxide. High conductivity (low resistance) of the electrolyte results in higher current ratings than the carbon-zinc cell. The alkaline cell is supposed to be a primary cell. However, I have been recharging them for years after finding out by experiment that it could be done. I charge these cells at 50mA constant current overnight. I first placed them inside a plastic garbage bin in case they exploded, but I have never had a misadventure with one yet. I have found that over the last few years, chargers have become available to recharge them! I guess they just did not want the public to know about it! These are the Copper Top and Energizer type batteries.

All chemical reactions are slowed by reduced temperature. Even at zero degrees centigrade, some cells give up the ghost. Alkaline cells have superior low-temperature characteristics compared to the older carbon zinc cell.

LEAD-ACID STORAGE BATTERY

The most widely used high capacity rechargeable battery is the lead-acid type. In automotive service, the battery is usually expected to discharge partially at a very high rate and then to be recharged promptly while the alternator is also carrying the electrical load. If the conventional car battery is allowed to discharge fully from its nominal 2.2 V per cell to 1. 75 V per cell, fewer than 50 charge-discharge cycles may be expected, with reduced storage capacity. The typical lead-acid car battery consists of 6 cells connected in series.

Many older car batteries could be physically sliced into six individual 2.2V cells. The plates in a lead-acid battery are lead peroxide for the positive and spongy lead for the negative. The electrolyte is dilute sulphuric acid.

SULPHATION

As a lead-acid cell discharges, lead sulphate is deposited onto the plates. This is commonly referred to as sulphation. Lead sulphate is a white powder often seen on the outside of old batteries or on the terminals. If lead-acid batteries are allowed to remain discharged for very long the plates will be covered in lead sulphate and the

capacity of the battery greatly reduced as the lead sulphate insulates the plates from having contact with the electrolyte.

Lead-acid batteries make a very cheap alternative to an expensive low voltage power supply. They are cheap to buy and can be trickle charged with an inexpensive charger.

Lead-acid batteries are also available with gelled electrolyte. Commonly called gel cells, these may be mounted in any position if sealed, but some vented types are position sensitive.

Lead-acid batteries with liquid electrolyte usually fall into one of three classes:

1. Conventional, with filling holes and vents to permit the addition of distilled water lost from evaporation or during high-rate charge or discharge.
2. Maintenance-free, from which gas may escape, but water cannot be added.
3. Sealed.

SPECIFIC GRAVITY

Specific gravity is the weight of a substance compared to the weight of water. An hydrometer (figure 16-3) can be used to draw some electrolyte from a lead-acid cell into a reservoir tube; a weighted float can be used to measure the specific gravity of the electrolyte. This reading gives a very good indication of the state of charge of a cell. In a fully charged cell, the specific gravity is in the vicinity of 1.280 (which means the density of the electrolyte is 1.280 times that of water). When the specific gravity is down to about 1.150, the cell is fully discharged. Water has a specific gravity of 1.0 so at an electrolyte level of 1.15, means we almost have water for an electrolyte.

Approximate State of charge	Average Specific Gravity	Open Circuit Voltage
100%	1.265	12.65V
75%	1.225	12.45V
50%	1.19	12.24V
25%	1.155	12.06V
0%	1.12	11.89V

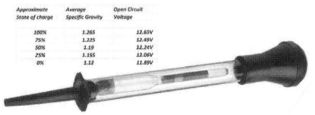

Figure 16-3

As you can imagine, when a lead-acid cell discharges, its electrolyte becomes weaker in sulphuric acid and approaches water. As a lead-acid cell discharges, the specific gravity or density of its electrolyte decreases (it turns more towards water).

WARNING

Lead-acid cells produce hydrogen and oxygen when being charged, these are highly flammable gases in combination with air. The bubbles in a liquid electrolyte are hydrogen. The oxygen remains mostly locked in the cell due to other chemical reactions. It is the combination of liberated hydrogen with air that is hazardous.

I have seen a telephone exchange levelled to the ground by the ignition of hydrogen gas from the charging cells. Neither hydrogen nor oxygen is flammable independently, but they are flammable when found in combination with air. Pure hydrogen will not burn by itself nor will oxygen. A spark in a container of hydrogen won't cause an explosion. Likewise, a spark in a container of pure oxygen may only form a bit of ozone. But when oxygen and hydrogen are combined, a vigorous reaction takes place when the mixture is ignited.

It is dangerous to allow salt water to get into lead acid cells; though it is a bit hard to avoid this in a sinking boat. Salt water in combination with sulphuric acid releases huge amounts of chlorine gas, a very heavy green gas which if inhaled causes severe burning of the respiratory system.

If a lead-acid cell is short circuited, enormous current will be drawn from the battery. It will overheat in seconds and explode spraying sulphuric acid everywhere. This sometimes unfortunately happens in car accidents.

NICKEL-CADMIUM BATTERY (NiCd)

NiCd batteries held the top spot as the battery of choice for use in portable equipment for decades. Today the NiCd has been displaced as best choice by the Lithium battery. The nickel-cadmium (NiCd), has a nominal voltage of 1.2 - 1.25 V per cell.

The NiCd cell has a positive electrode made of nickel and a cadmium negative electrode. The electrolyte is potassium hydroxide.

Carefully used, these are capable of 500 or more charge and discharge cycles. To ensure longer life, the NiCd battery should not be fully discharged. Where there is more than one cell in the battery, the most discharged cell may suffer polarity reversal, resulting in a short circuit, or seal rupture. All storage batteries have discharge limits and NiCd types should not be discharged to less than 1.0 V per cell.

Nickel cadmium cells are not limited to "D" cells and smaller sizes. They also are available in larger varieties ranging to mammoth 1000Ah units having carrying handles on the sides and caps on the top for adding water, similar to lead-acid types. These large cells are sold to the aircraft industry to start jet engines and to railroads for starting locomotive diesel engines. They also are used extensively for uninterruptible power supplies. Although expensive, they have very long life. Surplus cells are often available through surplus electronics dealers and these cells often have close to their full rated capacity.

Vented-cell batteries have and advantage of high discharge current to the point of full discharge. Also, cell reversal is not the problem that it is in the sealed cell since water lost through gas evolution can easily be

replaced. Simply remove the cap and add distilled water. By the way, tap water should never be added to either nickel-cadmium or lead-acid cells, since dissolved minerals in the water can hasten self-discharge and interfere with the electrochemical process.

114

NiCd batteries must be charged with a constant current source. A constant current source is one where the current through the cell does not change as the battery is charged. The rule of thumb is 1/10th the milliamp hour rating for 15 hours. So if a NiCd battery is rated at 500mA/hours, then it should be charged for about 15 hours at 50mA constant current if it is fully discharged. If the NiCd is only, 1/3 discharged then it should be charged for 5 hours.

LITHIUM CELLS

Like any other battery, a rechargeable lithium-ion battery is made of one or more power-generating compartments called cells. Each cell has essentially three components as all other chemical cells: a positive electrode, a negative electrode and an electrolyte in between them.

The positive electrode is typically made from a chemical compound called lithium cobalt oxide (LiCoO2) or, in newer batteries, from lithium iron phosphate (LiFePO4). The negative electrode is generally made from carbon (graphite).

The electrolyte varies quite a lot with many options available. Electrolytes comprise of lithium salts or organic solvents, such as ethylene-carbonate, dimethyl-carbonate and diethyl-carbonate.

All lithium-ion batteries work in broadly the same way. When the cell is charging up, the lithium-cobalt oxide, the positive electrode gives up some of its lithium ions, which move through the electrolyte to the negative graphite electrode and remain there. The battery takes in and stores energy during this process. When the battery is discharging, the lithium ions move back across the electrolyte to the positive electrode, producing the energy that powers the battery. In both cases, electrons flow in the opposite direction to the ions around the outer circuit. Electrons do not flow through the electrolyte: it's effectively an insulating barrier, so far as electrons are concerned.

The movement of ions (through the electrolyte) and electrons (around the external circuit, in the opposite direction) are interconnected processes and if either stops so does the other. If ions stop moving through the electrolyte because the battery completely discharges, electrons can't move through the outer circuit either- so you lose your power.

Lithium cells if overcharged or short circuited can cause an explosion. Large batteries usually have some electronic control and safety protection. Generally, lithium-ion batteries are more reliable than older technologies such as nickel-cadmium (NiCd) and don't suffer from a problem known as the "memory effect" (where NiCd batteries appear to become harder to charge unless they're discharged fully first). Since lithium-

ion batteries don't contain cadmium (a toxic, heavy metal), they are also (in theory, at least) better for the environment-although dumping any batteries (full of metals, plastics and other assorted chemicals) into landfills is never a good thing. Compared to heavy-duty rechargeable batteries (such as the lead-acid ones used to start cars), lithium-ion batteries are relatively light for the amount of energy they store.

CHARGING CELLS AND BATTERIES

RATE	TYPE	(C)	TIME	(C=mA/hour rating)
SLOW	NiCd, Lead Acid	0.1C	14hrs	Continuous or timed
RAPID	NiCd, NiMH,Li-ion	0.3-0.5C	3-6hrs	Sense voltage/temperature Timer recommended
FAST	NiCd, NiMH,Li-ion	1C	1hr	Safety same as rapid
SUPER FAST	NiCd, NiMH,Li-ion	1-10	10-60Min.	Special batteries and charging circuits. Monitor voltage, current and temperature

A charger for a lithium battery can output a constant voltage of 4.2V per cell in series. If a lithium cell does not receive more than 4.2V it can remain connected to the charger and will not over-charge.

CELLS IN SERIES AND PARALLEL

Cells can be connected in series, positive to negative to increase the output voltage. This method is common to all batteries. The small 9 volt transistor battery is 6 cells connected in series. The 6V lantern battery has 4 completely separate cells inside the case. If you get the opportunity, I suggest you break an old one open and have a look. It isn't too messy and you will find what looks like 4 oversized 'D' cells inside.

Batteries or cells can be connected in parallel to increase the current capacity. The batteries must be identical. A more desirable way to get greater capacity is to use a bigger battery in the first place.

It is very rare indeed to have batteries connected in a combination of series-parallel. This would only be done in specialist applications and is not recommended.

TESTING BATTERIES

The best way to test a lead-acid cell is by measuring the specific gravity of the electrolyte as already discussed. Smaller batteries and cells must be tested on a load. This means the equipment in which the battery is installed should be turned on and then the terminal voltage of the cell or battery measured. A fully discharged battery will still show its correct terminal voltage if it is not under load. It is important you understand this and the reasons why. As a battery ages its internal resistance increases.

A new battery will have an internal resistance, which is a small fraction of an ohm. The electrolyte between the plates forms part of the total circuit. If the battery is supplying 1A and the internal resistance of the battery is 0.1 ohms then a mere 0.1 volts will be dropped across the internal resistance, which is insignificant. As the battery discharges, the internal resistance will increase. However, no voltage is lost across the internal resistance unless the battery or cell is placed on a load. The load should be the nominal load that the battery is expected to deliver.

ELECTROLYTIC CORROSION

The use of dissimilar metals in engineering (say when building an antenna) is likely to cause considerable trouble due to electrolytic corrosion. Every metal has its own electro-potential and unless metals of similar potential are used the difference will cause corrosion at the point of contact even when dry. When moisture is present, this effect will be even more severe. If for any reason, dissimilar metals must be used then considerable care should be taken to exclude moisture, the corrosive effects of which will vary with atmospheric pollution.

Metals can be arranged in what is called the electrochemical series, as follows:

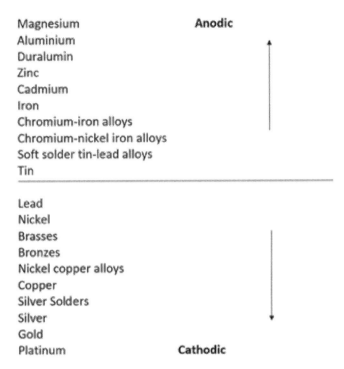

Metals from within the same group may be used together with little corrosion, but metals used from across the groups will suffer greatly from the voltaic corrosion effects. Also, since the above groups are arranged in order, the greater the spacing in the list, the greater will be the corrosion.

The metals of the lower group will corrode those in the upper portion. For example, brass or copper screws in aluminium will corrode the aluminium considerably, whereas with cadmium plated brass or copper screws there will be much less corrosion of the aluminium.

17 – Microphones

SOUND

Sound consists of small fluctuations in air pressure. We hear sound because these changes in air pressure produce fluctuating forces on our eardrums. You only have to place your hand in the front of your mouth while you are talking, or in front of a sound system, to feel the pressure waves.

Similarly, microphones respond to the changing forces on their components and produce electric currents that are effectively proportional to those pressure waves.

The best way of detecting sound is with a diaphragm ('dia'-'fram'). Our eardrums have a diaphragm. Just like the skin or membrane covering of a musical drum. When sound waves enter the ear, the force of the sound waves causes the eardrum to vibrate. Vibrations from the eardrum are transmitted by various structures in the ear to our brain for processing.

All microphones have a diaphragm, which usually consists of some type of taut material like plastic, paper, or even very thin aluminium. With all microphones, sound waves strike a diaphragm and cause it to vibrate. This vibration is then, by the mechanism of the microphone, made to make a voltage or current vary in sympathy with the sound waves. The voltage or current thus produced can be made to travel long distances (if necessary) along conductors and then we can recover the sound by converting current variations back into sound wave pressure variations.

THE CARBON MICROPHONE

Carbon microphones are no longer used, as they are low quality. However, a short discussion of the carbon microphone is a good introduction to all microphones. The carbon microphone was for many decades the only type of microphone used in telephones and two-way radios.

The principle of operation was both simple and ingenious. Sound waves strike the diaphragm and move it. The movement of the diaphragm causes a plunger or piston to move with it.
The plunger compresses and decompresses a chamber filled with carbon granules.

Now, carbon granules will conduct electric current. If a battery is connected to the microphone terminals, a current will flow.

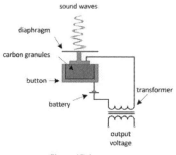

Figure 17-1

How much current flows will depend on the degree of compression of the carbon granules. If the sound waves force the diaphragm inwards, then the granules are compressed and their resistance decreases and the current increases. Similarly, if the sound amplitude (loudness) decreases, the tension of the diaphragm allows it to move outwards, the carbon granules decompress, the resistance increases and so the current decreases.

In other words, the resistance of the carbon microphone varies in sympathy with the sound waves that strike the diaphragm and this, in turn, causes a varying current in the circuit which is a good representation of the original sound wave.

Carbon microphones were used for many years. You may have seen people bang an older type telephone handset against something to get rid of the noise. This is because the carbon granules would cling together in older microphones and make a noise called 'frying' because the noise sounded like something frying in a pan.

THE DYNAMIC OR MOVING COIL MICROPHONE

In the dynamic microphone, the diaphragm is connected to a lightweight coil. Inside the coil is a permanent magnet. When sound strikes the diaphragm, the coil will move back and forth in the magnetic field of the permanent magnet. We have relative motion between a conductor and a magnetic field (Faraday's Law), so an emf or voltage is induced into the coil. This induced emf is a very good electrical representation of the original sound wave. A dynamic microphone produces its own output voltage and does not need a battery like the carbon microphone.

Dynamic microphones can be made in various impedances. (Impedance is the total opposition to current flow in an AC circuit). The dynamic microphone or some variation of it is widely used in telephones, radiocommunications and quality sound recording.

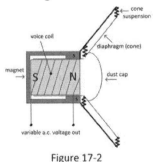

Figure 17-2

The speaker is essentially the same as the dynamic microphone, except the current carrying the sound to be recovered is fed through the coil, which creates a moving magnetic field around the coil. The interaction between the magnetic field of the coil and the field of the permanent magnet cause the diaphragm to move back and forth. The motion of the diaphragm recreates the original sound wave.

A SIMPLE TELEPHONE (FAR SOUND)

You do not need to know how a telephone works, however, looking at a simple telephone circuit will help to consolidate what you do need to know. In figure 17-3, we have a carbon microphone and a receiver in each handset.

A speaker is called a 'receiver' in telephony. A source of DC is applied to each microphone. The carbon microphones are coupled to the receiver and telephone line via a transformer. Sound on either carbon microphone will be converted to varying DC in the handset. The transformer will convert varying DC to AC and sound will be heard in both receivers.

The current contains all of the information present in the original sound wave (amplitude and frequency). Since the current flows through the receiver (most likely a moving coil), the diaphragm (cone) of both receivers will move back and forth with the same frequency and relative amplitude of the original sound wave and therefore recreate it.

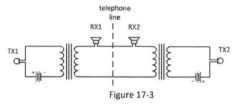

Figure 17-3

The distance between the microphone and the speaker can be a long way apart enabling voice communications over a long distance through wires. The modern landline telephone is really not much more than this in principle.

The carbon microphone that was used in telephones for many decades has been replaced by a variant of the moving coil type called a 'rocking-armature'. The rocking armature microphone is just a moving coil microphone with an inbuilt mechanical advantage between the diaphragm and the coil.

It seems simple does it not? Yet the impact of the telephone when it was first invented had a profound effect on the world. Of course, the telegraph came first. However, imagine how mysterious and wonderful it must have been for the first telephone users to hear and talk to someone thousands of miles away.

The word telephone comes from the Greek roots 'tele' ("far") and 'phone' ("sound"). Alexander Graham Bell invented the telephone (disputed), though many others improved on the invention. The U.S. patent granted to Bell in March 1876 (No. 174,465) for the development of a device to transmit speech sounds over electric wires, is often said to be the most valuable patent ever issued.

CONDENSER (CAPACITOR) MICROPHONE

When we learned about capacitance, we found that the capacitance of a capacitor was determined by the area of the plates, the type of dielectric and the distance between the plates.

Now, suppose a capacitor is connected to a source of emf and allowed to charge fully (at least 5-time constants). What would happen if we left the capacitor in the circuit and somehow altered its capacitance?

If we increased the capacitance of the capacitor in the circuit, it would be able to hold more charge. So increasing the capacitance would cause the capacitor to charge further and current would flow while the charging process was taking place. If we then decreased the capacitance, the capacitor would no longer be able to hold as much charge and it would discharge back into the source of the applied emf.

While it is discharging, current would again flow in the circuit.

In the capacitor microphone, that is what happens. One of the plates of the capacitor is made of very light material and is the diaphragm of the microphone. Sound waves striking the diaphragm (one plate) will cause it to vibrate in sympathy with the sound waves. The moving diaphragm (plate) will cause the capacitance to change. As the diaphragm moves in, the capacitance will increase and a charge current will occur. As the diaphragm moves out, the capacitance decreases and a discharge current occurs. Since the motion of the diaphragm and the capacitance are in sympathy with the sound waves, the charge and discharge currents are once again an electrical representation of the original sound wave.

The diaphragm of the microphone can be one of the plates of the capacitor microphone, or a diaphragm can be connected mechanically to make a capacitor plate move.

Like a carbon microphone, the capacitor microphone needs a source of emf, as it does not generate any of its own.

The term 'condenser' is an old term for 'capacitor'. When you see these microphones referred to, the words 'condenser' and 'capacitor' are often used interchangeably. I prefer 'capacitor microphone' but it seems for many the term 'condenser' has stuck from tradition. Capacitor microphones have the advantage of being very small, which makes them attractive for use in small equipment, such as handheld two-way radios (transceivers), small voice recorders and the like.

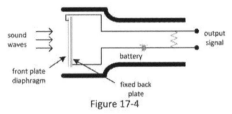

Figure 17-4

There is an another type of capacitor microphone called the 'electret microphone.' The principle of operation of these microphones is identical to that of a capacitor microphone. However, a special electret material is used as the dielectric in the capacitor. The electret microphone does not require a bias battery. An electret material is one that holds a charge for many years similar to the way a magnetic material can hold a magnetic field for a long time.

For exam purposes, you need only remember that an electret is the same as an ordinary capacitor microphone. However, it can be used without a bias voltage.

The dynamic and capacitor microphones are the most popular microphones in use today for radiocommunications.

CRYSTAL MICROPHONE

The Piezoelectric Effect (pee-zo-electric) microphone.

If a piece of quartz crystal is held between two flat metal plates and the plates are pressed together, a small emf will be developed between the plates as if the crystal became a small battery for an instant. How much emf is produced is proportional to the pressure applied. When the pressure on the plates is released the crystal springs back and an opposite polarity emf is produced on the plates. In this way, mechanical energy is converted into electrical energy by the crystal. Piezoelectricity means "pressure-electricity".

If an emf is applied to the plates of a crystal, the physical shape of the crystal will distort. If an opposite polarity emf is applied, the crystal will reverse its physical distortion. In this way, a quartz crystal converts electrical energy into mechanical energy and vice versa.

Some ceramic materials also exhibit a piezoelectric effect. These two reciprocity qualities of a quartz crystal and ceramics are known as the piezoelectric effect.

Well, if you are thinking ahead of me, here we have another way to make a microphone. After all, the function of a microphone is to convert mechanical energy (sound waves) into electrical energy and the piezoelectric effect does just this.

Sound waves striking the diaphragm cause varying pressure to be applied to the crystal, which in turn causes the microphone to produce an output voltage in sympathy with the sound waves. A crystal microphone does not require a battery. Like the dynamic microphone, it directly converts mechanical energy into electrical energy.

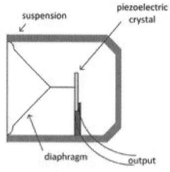

Figure 17-5

The crystal earpiece is the same principle used in reverse. Crystal microphones are not high quality, but they have the advantage of being small and inexpensive. The quartz crystal is not used in the form you may have seen them in nature. They are cut into thin slices and ground, like a gemstone is ground, to the exact size.

Well, that's about it for microphones. Recording or sound engineers use many speciality microphones. For example, special microphones that pick up sound from some directions and not others, etc. However, for voice communication, a simple dynamic or capacitor microphone does the job nicely.

MICROPHONE AMPLIFICATION

Because the output of a microphone is very low in electrical terms, in almost all applications, an amplifier called a microphone amplifier or pre-amplifier, is connected to the microphone to increase the level of the output signal. Many microphones come with a built in amplifier in the microphone housing.

Alexander Graham Bell (born March 3, 1847, Edinburgh. Died, August 2, 1922, Beinn Bhreagh, Cape Breton Island, Nova Scotia, Canada). Scottish-born American audiologist best known as the inventor of the telephone (1876). For two generations his family had been recognised as leading authorities in elocution and speech correction, with Alexander Melville Bell's Standard Elocutionist passing through nearly 200 editions in English. Young Bell and his two brothers were trained to continue the family profession. His early achievements on behalf of the deaf and his invention of the telephone before his 30th birthday,
bear testimony to the thoroughness of his training.

Antonio Meucci. There is much controversy over who really invented the telephone. There is strong evidence that an Italian immigrant to the United States by the name of Antonio Meucci invented the telephone before Bell and that his apparatus was misplaced or stolen. The evidence is so strong the US Congress passed a Bill naming Meucci as the inventor of the Telephone. A copy of this Bill can be easily found with an Internet search.

18 – Transformers

One of the most common devices used in electricity, electronics and radio, is the transformer. The name itself indicates that the device is used to transform, or change, something. The transformer may be used to step up or step down AC voltages, to change low voltage high current AC to high voltage low current AC, or vice versa, or to change the impedance of a circuit to some other impedance in order to transfer energy more efficiently from a source to a load.

A BUZZER

A buzzer has nothing to do with transformers. However, I am going to describe how a buzzer works as this will lead nicely to transformers.

Take a close look at the pictorial diagram of a buzzer in figure 18-1. It is so easy to make one of these on a piece of board. No need for you to make one but it would be good if you can picture how it is made and how it works. The electromagnet is just a soft iron bolt (any bolt would do) about 75mm long. Wind as many turns of thin single strand insulated wire on the bolt as you can get. The armature is just a strip of tin about 5mm wide and 75mm long. A nail holds it on the board on the left hand side. The contact on the right hand side is just another nail.

The strip of tin (the armature) forms part of the circuit for current flow. The circuit as it is now in figure 18-1 can conduct current. When the battery has been connected for a fraction of a second the magnetic field builds up around the electromagnet and the electromagnet pulls the armature towards it. When the armature moves towards the electromagnet, the circuit is broken and the current stops.

The magnetic field about the electromagnet collapses and the armature springs back to the contact. Current can now flow again. This cycle is repeated over and over. The result is that the motion of the armature hitting the electromagnet makes a buzzing sound. We have a buzzer! If the armature was extended with a rod and on the end of that rod we had a small weight (hammer) that could strike a gong; we would have an electric bell. What type of current flows in this circuit?

I hope you agree it is pulsating DC, or DC which is being turned ON and OFF rapidly. Now think about the magnetic field around the electromagnet. It is a pulsating magnetic field, or perhaps more importantly, it is a magnetic field which is continually varying. It never stays still as it is either expanding when the current is ON or collapsing when the current is OFF.

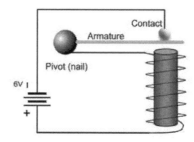

Figure 18-1

Let's make a modification to our simple buzzer circuit to demonstrate the action of a transformer. Refer to figure 18-2.

All we have done is wind a second coil of wire, either on top of or next to, the electromagnet using insulated wire and connected it to a voltmeter. The two coils (inductors) are insulated from each other.

Now, do you remember when we discussed inductance and alternating current? We talked about Faraday's law of magnetic induction.

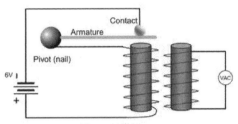

Figure 18-2

Faraday's law states: When relative motion exists between a conductor and a magnetic field an emf is induced into the conductor.

Is there relative motion between the conductors on the second coil and a magnetic field? Yes, there is. An emf (voltage) is induced into the second conductor and this will be shown on the voltmeter. We have created a simple, though very inefficient transformer. Generally, a transformer has two windings. The primary winding is supplied with some type of current that will cause the magnetic field around it to vary. In our experiment, the primary winding is the coil of wire around the electromagnet. The other winding of a transformer is called the secondary winding.

A transformer is basically two or more windings (coils, inductors) wound on a common core or former. The primary winding is fed with some type of varying current so that a moving magnetic field is created around it. The moving magnetic field created by the primary winding causes an induced emf into the secondary winding.

For a transformer to work the primary winding must have a current through it that produces a varying magnetic field. This is virtually any current other than DC, such as AC, pulsating DC, varying DC and the like. The secondary voltage (type) will always be AC.

What we need to learn are the uses for a transformer and how we can determine the amount of induced voltage into the secondary.

Obviously, we could reason that the amount of induced voltage into the secondary would have something to do with the voltage on the primary and the ratio of the number of turns on the primary and secondary. We will discuss this relationship in more detail shortly.

Transformers do not have to have a magnetic material such as iron for a core. The core can be air. The basic function of a transformer is to step voltage up or to step voltage down, which is also a way of matching unequal impedances.

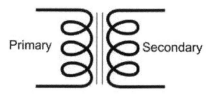

Figure 18-3

The schematic symbol for a transformer is shown in figure 18-3 (excluding the writing which I have added). The two vertical lines between the two windings indicate that this transformer has a laminated iron core. If no lines were shown, it would be an air core transformer. You do not know what the number of conductor turns are on the primary or secondary from the symbol. More important is the ratio of the number of turns on the primary to the secondary. This ratio is called the turns ratio. If there are more turns on the secondary than the primary, then voltages are stepped up. Conversely, fewer turns on the secondary and voltage is stepped down. The symbol may show you if the transformer is step up or step down but it will not give the turns ratio. The voltage ratio is the same as the turns ratio. If voltages for primary and secondary are given, then you know the turns ratio since voltage and turns ratio are the same.

TRANSFORMER CONSTRUCTION

Figure 18-4 shows one construction method for a laminated iron core transformer. Here the primary and secondary are wound separately around a laminated iron core.

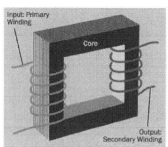

Figure 18-4

LAMINATED IRON CORE

Winding the primary and secondary on an iron core improves the efficiency of the transformer since the iron core concentrates the magnetic lines of force.

The problem with iron cores is that the core is also a conductor. Think back to Faraday's law again. The core of a transformer (iron) is in a moving magnetic field. This means that small currents will be induced into the core. These unwanted core currents are called Eddy Currents.

These currents are given this name because they flow in circles in the iron core much like eddies of water around a propeller.

Eddy currents are undesirable because of $P=I^2R$ losses. What this means is that current flowing through any resistance produces heat. This is fine if you want to make heat. However, the function of a transformer is not to make heat but to transform from one voltage to another. By the way, you may hear the term "I squared R losses" used. This means the same thing, though "I squared R losses" can occur other ways besides eddy currents.

What is the product of the current and voltage in a circuit? Power= E x I. A transformer should ideally have no losses. This means the power in the primary circuit should be equal to the power in the secondary circuit. The power in each circuit is the product of the voltage and current in the respective circuit.

In practical transformers, particularly iron core ones, the major form of loss of power is due to eddy currents, also known as I squared R losses.

If we could increase the electrical resistance of the core without changing its magnetic properties, we would reduce the magnitude of the eddy currents. When we discussed the properties that determine the resistance of a conductor, one of these properties was the cross-sectional area. The smaller the cross sectional area the greater the resistance. This is what we are doing when we laminate the iron core.

Imagine taking a solid iron core and slicing it into lots of thin sheets. Each sheet (lamination) will have a higher resistance to current flow than the solid iron core. We now spray each sheet with an insulating lacquer and bolt or glue the entire iron core back together again. We now have a laminated iron core, which will have less power lost in it due to eddy currents.

Eddy currents are not a problem if the core of a transformer is made of air or some other non-conductive material.

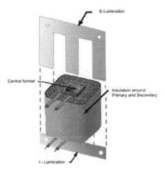

Figure 18-5

Another construction method for a transformer is shown in the diagram of figure 18-5. Eddy currents are reduced in amplitude by laminating the core. I have shown one E-lamination and one I-lamination. Imagine 50 or more of such laminations. Insulated from each other, then bolted together to form a solid laminated iron core.

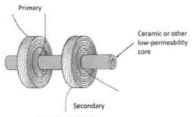

Figure 18-6

The type of transformer shown in figure 18-6 is used at high frequencies. Iron core transformers are not used at radio frequencies, as the eddy current losses become too great.

HYSTERESIS - LOSSES

If iron is in an unmagnetised state, its magnetic domains are not arranged in any particular manner. The domains are randomly oriented. When a magnetising force is applied to them, the domains rotate into a position in line with the magnetising force. If the magnetising force is reversed, the domains must rotate to the opposite position. In rotating from one alignment to the opposite, the domains must overcome a frictional hysteresis, or a resisting effect, in the core. In some materials the resisting effect is small, in others it is appreciable. Due to this effect, we lose some energy as heat. Heat is nearly always lost energy. This loss is called **hysteresis loss**.

Hysteresis occurs in iron cores of transformers. As frequency is increased, the alternating magnetising force will no longer be able to magnetise the core completely in either direction. Before the core becomes fully magnetised in one direction, the opposite magnetising force will begin to be applied and start to reverse the rotation of the domains. The higher the frequency, the less fully the core magnetises.

Transformers operated on low-frequency AC may not have much hysteresis loss, but the same cores used with a higher frequency have more hysteresis and are less efficient.

COPPER LOSS

Iron core transformers are subject not only to eddy current and hysteresis losses in the core but also to a copper loss which occurs as heat lost in the resistance of the copper wire making up the windings. The current flowing through whatever resistance exists in these windings produces heat. The heat in either winding, in watts, can be found by the power formula $P = I^2R$. For this reason, the copper loss is also known as the I^2R loss. The heavier the load on the transformer (the more current that is made to flow through the primary and secondary), the greater the copper loss.

With one layer of wire wound over another in a transformer, there is a greater tendency for the heat to remain in the wires than if the wires were separated and air cooled. Increased temperature causes increased resistance of a copper wire. As a result, it becomes necessary to use heavier wire to reduce resistance and heat loss in transformers than would be required for an equivalent current value if the wire was exposed to air during operation.

EXTERNAL INDUCTION LOSS

Another loss in a transformer is due to external induction. Lines of force expanding outward from the transformer core may induce voltages and therefore currents into outside conducting bodies. These currents flowing through any resistance in an

outside body will produce a heating of the external resistance. The power lost in heating these outside circuits represents a power loss to the transformer, since the power is not delivered to the transformer's secondary circuit. In a well designed transformer, the amount of power lost in this fashion is usually small. The efficiency of a well designed transformer is around 98%.

THE VOLTAGE RATIO OF TRANSFORMERS

One of the main uses of transformers is to step up a low voltage AC to a higher voltage. This can be accomplished by having more turns on the secondary than on the primary.

The turns ratio of a transformer is the ratio of the number of turns on the primary to the number of turns on the secondary.

If a transformer has an equal number of turns on the primary and secondary (it happens), then the turns ratio is 1:1.

If a transformer has 100 turns on the primary and 10 turns on the secondary, then the turns ratio is 100:10 which is the same as 10:1.

If a transformer has 300 turns on the primary and 900 turns on the secondary, then the turns ratio is 300:900 which is the same as 1:3.

The shorthand for primary turns is usually Np and for secondary turns Ns. The voltage of the primary is usually designated Ep and for the secondary Es.

The turns ratio of a transformer is the same as the voltage ratio. In other words:

Np/Ns = Ep/Es

A Worked Example.

A transformer has 100 turns on the primary and 10 turns on the secondary. If 500 volts AC is applied to the primary what will be the secondary voltage?

Np/Ns = Ep/Es

We could transpose this equation for Es but for exam purposes an easier approach is to see that the turns ratio of the transformer Np/Ns is 100/10 or 10:1. For every 10 volts on the primary, there will be 1 volt on the secondary. Therefore, the secondary voltage will be 500/10 or 50 volts.

This is a step-down transformer because the secondary voltage is lower than the primary voltage. A step up transformer is where the secondary voltage is higher than the primary voltage

The turns ratio is the same as the voltage ratio.

A lot of learners get confused, especially with the step up transformers. You can make a transformer to convert 10 volts AC to 10,000 volts AC. At first sight, it may appear that you are getting something for nothing - no such luck.

The power in the primary circuit is equal to the power in the secondary circuit (disregarding losses).

No power is gained in a transformer. If a transformer steps the voltage up, there is a corresponding decrease in secondary current. So we may get more secondary voltage, but at the sacrifice of the secondary current.

Take an arc welder. This is basically a huge step-down transformer. 240 volts AC to the primary and only a few volts on the secondary, but a huge amount of current (60-100 amps or more) for melting metals is available in the secondary circuit.

TRANSFORMER ISOLATION

One major advantage of transformers is the safety isolation they can provide. Small battery eliminators are really just small transformers. The primary connects to 240 VAC and the secondary may be 12 VAC (or converted to DC).

Because there is no metallic contact between the secondary and the primary, it means the user of the low voltage equipment is isolated from the mains power, providing a significant safety advantage.

If you build an interface between electrical equipment and a computer for software control, you will most likely use some form of isolation. There are several methods, but one method is to use an audio isolation transformer. This will prevent DC levels in your transceiver affecting DC levels in the computer sound card.

AUTO TRANSFORMERS

An auto transformer consists of a single winding with one or more taps on it, as shown in figure 18-7.

Auto transformers are not "automatic". The name "auto" comes from their use as a car (auto) ignition coil.

Everything we have discussed about other transformers (mutually coupled transformers) applies to autotransformers.

Auto transformers used on high voltages are dangerous, as they provide no electrical isolation between the user and the supply. These transformers are not commonly used in radio and communications anymore. Auto transformers have the advantage of requiring less copper, therefore less weight and cost. However, in nearly all applications today better options are available than using auto transformers.

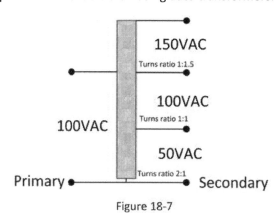

Figure 18-7

We have not discussed the purpose of impedance matching yet and I am not going to now. Suffice to say, in electronics; it is important to connect components together which are of the same impedance (impedance is the total opposition to current flow in an AC circuit).

For example, your sound system may specify that you use 8-ohm speakers. If you use speakers other than 8 ohms on your sound system, you may damage the sound system or at least get reduced quality sound.

A transformer could be used to connect any impedance speaker to a sound system. The impedance ratio of a transformer is related to the turns ratio. The relationship between turns ratio and impedance ratio is given by the equation:

$$\frac{N_p}{N_s} = \sqrt{\frac{Z_p}{Z_s}}$$

Z_p = impedance looking into the primary terminals from the power source.
Z_s = impedance of load connected to secondary and
N_p / N_s = turns ratio, primary to secondary.

This equation "says" *the turns ratio is equal to the square root of the impedance ratio and the impedance ratio is equal to the square of the turns ratio.*

Example.

A 100Ω source is connected to a 25Ω load using a transformer. What is the turns ratio of the transformer? Well, the impedance ratio is given as 100:25 which is 4:1 - this is the impedance ratio. Look at the equation at the bottom of page 161. The turns ratio is on the left and the impedance ratio on the right so one equals the other except there is a squareroot sign on the right-hand side. So turns ratio Np/Ns equals the square root of the impedance ratio. The square root of 4:1 (just take the square root of each number 4 and 1) is 2:1.

So to match a 100Ω source to a 25Ω load we would use a transformer with a 2:1 turns ratio.

How is a transformer able to match impedance?

Suppose we had a transformer with 2400 turns on its primary and 150 turns on its secondary. The turns ratio of this transformer is 2400 / 150 or 16:1.

The turns ratio is 16:1. Let's see if we can work out what the impedance ratio would be.

For the sake of this exercise let's apply a primary voltage of 240 volts. Since the voltage ratio is equal to the turns ratio, the secondary voltage must be 15 volts.

The turns ratio is 16:1 and this gives us the same voltage ratio, that is, 240/15 = 16:1.

Now for calculation purposes, I have to nominate a maximum secondary current that the transformer can handle. This value is arbitrary and any value can be chosen for our purposes. The maximum secondary current I will choose is 1 ampere.

This transformer is shown in figure 18-8.

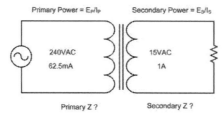

Figure 18-8

Since we know the secondary current and voltage (for full load) we can work out the secondary power:

$P_{secondary} = E_{secondary} \times I_{secondary} = 15 \times 1 = 15$ watts.

Since the power in the primary is equal to the power in the secondary (neglecting losses) we can, knowing the primary power is 15 watts, work out the primary current:

$I_{primary} = P_{primary} / E_{primary} = 15/240 = 62.5$ mA.

Where is all this taking us? Well, impedance is the ratio of voltage to current:

Z = E/I.

We know what the current and voltage is in the primary and secondary circuit, what is the primary and secondary impedance?

Impedance of the primary= $E_{primary} / I_{primary}$= 240volts / 62.5mA = **3840Ω**.
 Impedance of the secondary= $E_{secondary}/I_{secondary}$= 15/1 = **15Ω**.

In other words, the impedance ratio of this transformer is 3840:15 or 256:1. I have gone about it the long way just to show you how it works.

The rules you need to remember:

(a) The turns ratio is equal to the square root of the impedance ratio.
(b) The impedance ratio is equal to the square of the turns ratio. Np/ Ns = Square root of (Zp/Zs)

USING THE FORMULA

Let's work out the impedance ratio from the formula. From the last example, we know the voltage ratio is 240:15 which is the same as the turns ratio Np/Ns.

Np/ Ns = square root of (Zp/Zs) (remember Zp/Zs is the impedance ratio)
240 / 15 = square root of (Zp/Zs)

132

16 = square root of (Zp/Zs)

We do need to get rid of the "square root" on the right-hand side. To do this we "square" both sides and we are left with:
16 x 16 = Zp/Zs
or
Impedance ratio= Zp/Zs = 16 x 16 or 256 : 1

This agrees with our earlier calculation.

If you have trouble using this equation, ask your trainer for help. Many people do have some trouble, so don't be afraid to ask.

The type of question you need to answer in Advanced exams is something like this. What is the turns ratio of a transformer that is required to match a 600Ω source to a 50 ohm load?

Zp = 600Ω; Zs = 50Ω

Turns ratio= Np/ Ns = square root of (600/50) = 3.464:1.

For every turn on the secondary, there will be 3.464 turns on the primary.

Another.

A transformer is used to match a 300Ω source to a 50Ω load. What is the turns ratio of the transformer?

Using:
$$\frac{N_p}{N_s} = \sqrt{\frac{Z_p}{Z_s}}$$

The turns ratio (Np/Ns) equals the square root of the impedance ratio. Square root of (300/50) = 2.45:1

and·

The impedance ratio of a transformer is 10:1 what is the turns ratio?
The turns ratio is the square of the impedance ratio 10^2 = 100:1

The type of transformer that we have been discussing here is the mutually (magnetic) coupled transformer and can be used for impedance matching, however, their bandwidth is very narrow. A good example might be an RF auto transformer in the boot of a car to match a mobile vertical whip. This will work fine but because such a transformer will have a narrow bandwidth the usual thing is to have many tappings on it and when changing frequency even slightly on a short whip you might have to readjust the impedance match.

A better broad impedance match is the transmission line balun (to be discussed in transmission lines). You should know that while these may look similar to mutually coupled transformers, they have nothing in common at all. Transmission line baluns have no turns ratio, no primary or secondary, they do not work on the principle of Faraday's Law. Apply nothing you have learned in this chapter to transmission line baluns.

19 - Resistive, Inductive & Capacitive Circuits

The purpose of this chapter is to understand the combination of resistive, inductive and capacitive circuits and the concepts of impedance, quality factor or 'Q' and resonant circuits.

Impedance (symbol [Z])

Impedance is the opposition to current flow in an AC circuit, measured in ohms. There is no inductive or capacitive reactance in a DC circuit. However, in an AC circuit, there is resistance, inductive and capacitive reactance. All of these oppose current flow and it is this combined opposition that we call impedance.

Since impedance includes capacitive and inductive reactance it is frequency dependent. Resistance is not frequency dependent.

Impedance cannot, except in the case where reactances are equal and opposite, be expressed as a single number. There are two standard methods of expressing impedance.

1. The **rectangular form** and;

2. The **polar form**

RECTANGULAR FORM

When we discussed capacitive and inductive reactance, we talked about how the current leads voltage in a capacitive circuit, and current lags voltage in an inductive circuit.

When the reactances are equal ($X_L=X_C$) then there is no lead or lag. The lead effect of X_C and the lag effect of X_L cancels. We have no lead or lag even though we still have X_L and X_C in the circuit.

When the reactances are equal we have a special case called *resonance* and the circuit is *resistive*. There is no lead or lag. The current and voltage are in phase just as if we had a purely resistive circuit.

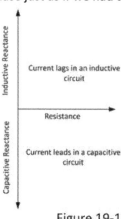

Figure 19-1

Notice how we show resistance as starting from an origin and moving horizontally to the right. In a purely inductive circuit the current lags the voltage and we show this as an upward vector from the origin. Similarly a capacitive circuit causes the current to lead the voltage so we show this as a downward vector from the origin.

The reactance vectors are opposed at 180 degree so that the larger one (if there is a larger one) cancels the smaller one. If they are equal there is complete cancellation and the net reactance is zero. Suppose we had a RLC circuit comprising of 100Ω of R and 100Ω of X_C and 100Ω of X_L. We could plot these opposition vectors as shown in figure 19-2.

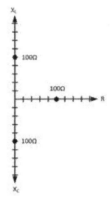

Figure 19-2

Since the X_L and X_C vectors are opposed to each other and the same length (100Ω) they will cancel and the net reactance is zero and the only opposition left is the resistance (R).

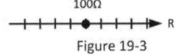

Figure 19-3

If we did have a RLC circuit comprising of R, X_L and X_C equal to 100Ω each then the impedance of that circuit would be 100Ω. However just saying the impedance is 100Ω does not really describe the circuit in full. There are two reactances there that are left out if we just say the impedance is 100Ω.

In mathematics R is said to be on the number line just like any real number. The reactances are ninety degrees off the number line and opposing. For naming sake we call the inductive reactance number line +j and the capacitive number line -j. In mathematics anything on the -j and +j number line are called IMAGINARY numbers! The resistive number line is called REAL.

These terms REAL and IMAGINARY just describe to us the way impedance is plotted in figure 19-2. Impedance written in the form R+/-JΩ is called the **rectangular form**. This will become more obvious shortly.

The impedance shown on figure 19-2 would be written down as 100j0Ω. This tells us the circuit is equal to 100Ω of resistance and the reactances cancel to give j0Ω.

Suppose we had the RLC circuit shown in figure 19-4.

135

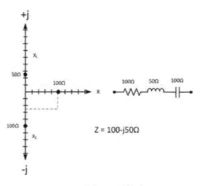

Figure 19-4

The capacitive reactance uses 50 of its ohms to fully cancel the inductive reactance of 50 ohms leaving a net capacitive reactance of 50 ohms. In rectangular form the net impedance is written as $100-j50\Omega$

In figure 19-5 the inductive reactance is now 150Ω and the capacitive reactance 100Ω. Now the inductive effects over power the capacitive reactance. The net reactance (X_L-X_C) is 50Ω of X_L and the impedance of the series RLC circuit is now $100j50\Omega$. By convention we do not write $+j50$. We drop the"+" and just write "j".

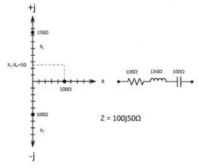

Figure 19-5

I think by now you have a good understanding of using rectangular notation for describing an impedance.

POLAR FORM

Take a look at figure 19-6. Here we have a resistance of 150Ω and our net reactance is equal to 150Ω and is inductive. In rectangular form we would say this is $150j150\Omega$

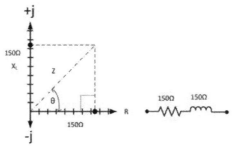

Figure 19-6

136

The other method of describing impedance is the polar form. In figure 19-6 notice we have drawn a right angle triangle the base of which is R and the height is X_L. The length of the hypotenuse is not known and neither is the angle θ. In polar form "Z" is called the magnitude of the impedance and θ the (phase) angle. What we need to do is use trigonometry to calculate Z and the angle θ. You probably can guess that the angle θ will be 45 degrees.

We can use Pythagoras to find Z and angle θ will be given by the trigonometric function Tangent.

$$Z_{mag} = \sqrt{R^2 + Xc^2} \qquad \tan\theta = \frac{X_L}{R}$$
$$Z_{mag} = \sqrt{150^2 + 150^2} \qquad \tan\theta = \frac{150}{150}$$
$$Z_{mag} = 212\Omega \qquad \tan\theta = 1$$
$$\arctan 1 = 45°$$

A series circuit comprising of 150Ω of "R" and 150Ω of "X_L" has an impedance in rectangular form of 150J150Ω and in polar form 212Ω/45 degrees.

In rectangular form R+/-JX R is called the **real** and X the **imaginary**. In polar form the impedance is referred to as **magnitude** and **angle**.

Let's do one more with some not so predictable figures.

With reference to figure 19-7 express the impedance of the circuit in both rectangular and polar forms.

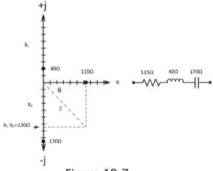

Figure 19-7

Rectangular:

The resistance is 115Ω and the net reactance (X_L-X_C) is 130Ω so the impedance of this circuit is 115-J130Ω.

Polar:

The magnitude of the impedance is; square root ($115^2 + 130^2$) = 173.5Ω.

The phase angle is arctan (X_C/R) = arctan(130/115) = 48.5 degrees.
The impedance in polar then is 173.5Ω angle 48.5 degrees.

Numbers expressed in polar or rectangular form are called "Complex Numbers". For finding the impedance of more complex circuits with series and parallel combinations the same rules apply as with resistors. However adding, subtracting, multiplying and dividing complex numbers has specific rules.

THE IMPEDANCE OF A PARALLEL LR CIRCUIT

A parallel circuit comprises of a 30Ω inductor in parallel with a 50Ω resistor. What is the impedance and phase angle of the circuit in polar form.

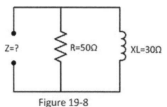

Figure 19-8

The equation and working out to find the impedance of a reactance in parallel with a resistance is:

$$Z_{mag} = \frac{RX}{\sqrt{R^2 + X^2}}$$

$$Z_{mag} = \frac{50 \times 30}{\sqrt{50^2 + 30^2}}$$

$$\Theta = \arctan(R/X)$$

$$Z_{mag} = \frac{1500}{58.3}$$

$$\Theta = \arctan(50/30)$$

$$Z_{mag} = 25.7\Omega \qquad \Theta = +59°$$

This same method can be used for a parallel RC circuit. If a parallel circuit contains many LCR branches then these can be simplified to a single resistor and reactance. Susceptance is the reciprocal of reactance in siemens. Admittance is the reciprocal of impedance in siemens. Conductance is the reciprocal of resistance in siemens.

Recall when we found the resistance of a number of resistors in parallel? We used this equation:

$$R_t = \frac{1}{1/R_1 + 1/R_2 + 1/R_3}$$

This equation says that the total resistance of resistors in parallel is the reciprocal of the sum of their reciprocals. You may not have realised it at the time but what you are doing here is converting the resistances to conductance. Adding the conductances and then finding the reciprocal of the sum to convert your answer from conductance back to resistance.

This same method can be used for any parallel circuit with any combination of reactances. The reactances must be treated as complex numbers and the rules for complex operations must be applied. Many scientific calculators will perform complex number operations and there are also many online calculators available.

RESONANCE.

Resonant circuits have some very useful properties in electronics and radiocommunications, however, let's talk a bit about what resonance is.

Electrical and mechanical resonance are very similar. A taut string, like that on a guitar, if plucked, will resonate at a particular frequency. The length of the string determines the frequency of resonance.

A metal tuning fork is designed to resonate at a particular frequency. When you strike a tuning fork, it will vibrate at the same single frequency every time, because the tuning fork is resonant at that frequency.

If you hit the rim of a wineglass with a hard object it will "ring" and it will always ring at the same frequency because it is resonant at that frequency.

In all the cases above, we had to deliver some energy to the object (guitar string, tuning fork or wineglass) to get it to resonate.

Have you ever been listening to loud music and at certain times the window may rattle, or an object on a table may start to move (vibrate with the music)? The window will start to vibrate if sound waves at its resonant frequency strike it. Since music contains many frequencies, the window will only vibrate when its resonant frequency is present in the music.

HOW DOES A SINGER BREAK A GLASS?

It is true that a singer provided they can reach the right note can break a wine glass. A wine glass has a frequency at which it will resonate. Making a loud noise will not break a wine glass. However, if a singer reaches and sustains a note which is equal to the resonant frequency of the wine glass, then the wine glass will absorb energy from the sound wave and it will begin to resonate (vibrate) and if the amplitude (loudness) is enough, it will shatter.

![Figure 19-9]

Figure 19-9

The wineglass absorbs energy on its resonant frequency. All objects, even electrical circuits, will absorb energy at their resonant frequency.

TUNING FORK EXAMPLE

Suppose we had two tuning forks designed to create the same note (frequency) and we struck one fork to make it resonate. Then place the two tuning forks say 50mm apart, the tuning fork that you did not strike would begin to resonate. The second tuning fork has absorbed sound wave energy from the first tuning fork and starts to vibrate in sympathy.

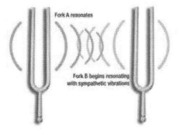

Figure 19-10

In electronics, we don't say something vibrates. We say it oscillates. A mechanic might say that guitar string, when plucked, vibrates at 800 times per second. A musician would say it makes such and such a note. An electronics person will say it oscillates at 800 hertz.

DAMPED OSCILLATIONS

If a guitar string or tuning fork is given energy, we have learned that it will oscillate at its resonant frequency. All things being perfect, the wave produced will be the shape of a sinewave. Now what happens to the sound wave emitted from a guitar string or whatever, after the initial pulse of energy is given to it?

The guitar string will oscillate vigorously at first and then slowly decrease in amplitude until it no longer makes any sound. Importantly - the frequency stays the same, but the amplitude dies down to nothing. This is a damped oscillation.

CONTINUOUS OSCILLATION

A swing (perhaps with a child on it, though the child is not an essential ingredient) will oscillate as it is a pendulum. We know that a pendulum will move back and forth at the same frequency (or period) every time. That is a fact because we use pendulums to create clocks. The amplitude of the swing of the pendulum will slowly dampen, but the frequency will remain the same.

Back to the child on the swing. How do we keep the rhythm going (oscillating)? Well, we have to keep giving it energy. When do you give it energy? Well if you are a good swing pusher, you know to give the extra push to sustain the motion of the swing when it is at the top of its cycle (about to change direction). You do not have to push hard to sustain the swing's oscillations. Once the swing is going, you only need to keep supplying enough energy to overcome the losses in the swing system. It is the losses in the swing, which slow it down. The losses in a swing are from friction.

Everything we have just discussed about resonance in mechanical systems (swings, guitars and the like) are equally applicable to resonant electronic circuits.

LC RESONANT CIRCUITS

Not every LC (inductive and capacitive) circuit is resonant. A resonant circuit is only resonant if we use it as a resonant circuit. Though every LC circuit will have a resonant frequency, we don't have to use it as a resonant circuit. We are going to talk about LC circuits that we want to resonate.

If we connect a capacitor and an inductor in parallel as shown in figure 19-11, nothing happens, as there is no energy in the circuit. This is just like the child's swing before we push it.

Now suppose we give the parallel LC circuit some energy. There are many ways we can give this circuit a pulse of energy. One way would be to expose the inductor to a moving magnetic field (Faraday's Law of Induction). We could disconnect the capacitor, charge it then connect it back to the inductor.

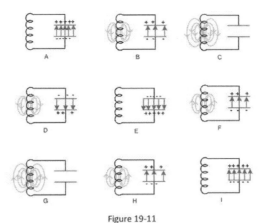

Figure 19-11

The latter is what I have done in figure 19-11. I have charged the capacitor and then connected it back in parallel with the inductor. What happens?

Look at 19-11A. We have given the capacitor some energy and all of that energy is stored in the electric field of the capacitor. The capacitor will begin to discharge through the inductor. As it does so, a magnetic field will be created around the inductor; 19-11B. Eventually, the capacitor will be discharged; 19-11C.

Now the energy has moved from the electric field of the capacitor into the magnetic field of the inductor. When the capacitor stops creating current in the circuit the magnetic field around the inductor cannot stay there, as it has no current to support it.

The magnetic field about the inductor will now begin to collapse into the inductor. In doing so, there will be relative motion between the inductor and the magnetic field and an emf will be induced into the inductor.

This emf will cause a current to flow in the opposite direction and will cause the capacitor to charge; 19-11D. Eventually, all the energy from the magnetic field will have returned to the circuit and the capacitor will be charged (storing energy in its electric field) 19-11E.

The capacitor will begin to discharge through the inductor again and so on.

This circuit is oscillating. The shape of the voltage produced is a sinewave. These oscillations would go on forever except, just like the swing, there are losses in the circuit, which dissipate energy. The losses are the resistance in the wire of the inductor, the circuit and the plates of the capacitor. There is also a little energy lost in the dielectric of the capacitor. The oscillation of this circuit then is a damped oscillation.

If we wanted a continuous oscillation, we would have to supply the circuit above with energy to overcome the losses. Such a circuit is called an oscillator.

We know what moves with mechanical resonance as we can see or feel it. The guitar string vibrates, the tuning fork vibrates back and forth. What does the moving with electrical resonance? It is the electric and magnetic fields that move. The capacitor dumps its electric field energy into the magnetic field of the inductor; then the inductor dumps its magnetic field energy back into the electric field of the capacitor. This happens at a certain rate and that rate is known as the resonant frequency of the circuit.

It should be intuitively obvious that if we had small Ls and Cs that could not hold much energy that the rate of dumping to each other would be faster - a higher resonant frequency. If we had large Ls and Cs then they could hold a significant amount of energy; then the rate of dumping would be much slower - a lower resonant frequency.

RESONANT FREQUENCY

If you thought about the circuit we have just described, you could probably guess what determines the resonant frequency or frequency of oscillation. Imagine if the capacitor was huge and took ages to discharge through the inductor. Likewise, if the inductor is large. You would expect the oscillations or resonant frequency to be low. You would be right. Large values of L and C resonate at a low frequency whereas small values of L and C resonate at a higher frequency.

To work out the exact resonant frequency (fr) of a tuned circuit we can use the following important equation:

$$fr = \frac{1}{2\Pi\sqrt{LC}}$$

The resonant frequency of a tuned circuit is inversely proportional to the square root of the product of the inductance and capacitance. The equation gives the resonant frequency in Hertz and L and C must be entered in Henrys and Farads.

You may have to use this equation in an Advanced exam. You may be asked to identify the equation from among others. You must know how to increase or decrease the resonant frequency of a tuned circuit.

The same equation is also used to find the resonant frequency of a series LC circuit. The resonance equation is one of the most important fundamental equations used in electronics. Now you might be wondering just how this equation is derived. Recall we said resonance occurs when $X_L = X_C$? This is equivalent to saying resonance happens when: -

$$2\pi f L = \frac{1}{2\pi f C}$$

$$\frac{2\pi C}{1} \times 2\pi f L = \frac{1}{2\pi f C} \frac{2\pi C}{1} \qquad \text{Multiply both sides by 2}\pi\text{C}$$

$$4\pi^2 f LC = \frac{1}{f} \qquad \text{Simplify}$$

$$\frac{f}{1} = \frac{1}{4\pi^2 f LC} \qquad \text{Rotate and invert}$$

$$f^2 = \frac{1}{4\pi^2 LC} \qquad \text{Multiply both sides f and simplify}$$

$$f = \frac{1}{2\pi\sqrt{LC}} \qquad \text{The resonance equation}$$

USING THE RESONANCE EQUATION:

$$fr = \frac{1}{2\Pi\sqrt{LC}}$$

A tuned circuit (series or parallel) consists of an inductor of 100 microhenries and a capacitor of 250 picofarads. On what frequency will it resonate?

100 microhenries = 100 x 10^{-6} henries
250 picofarads= 250 x 10^{-12} farads

Inserting these values into the equation we get:

Resonant frequency (fr) = 1 / (2 x π x square root (100 x 10^{-6} X 250 x 10^{-12}))
Resonant frequency= 1.007 megahertz.

THE 'Q' OF A RESONANT CIRCUIT

The "Q" is a term used for the 'quality' of a tuned circuit. We want as high a Q as possible. The more losses a tuned circuit has, the lower the Q. The fewer losses a tuned circuit has, the higher the Q. The equation below defines what Q is:

$$Q = \frac{Energy\ Stored}{Energy\ Lost\ Per\ Cycle}$$

You can see that Q is a measure of the losses in a tuned circuit or a lone inductor or capacitor. If a tuned circuit had no losses the Q would be infinite. There are many equations for calculating Q. However the one that is the most practical is:

$$Q = \frac{X_L}{R}$$

This is the equation you would use for an RLC circuit. You might think "what about capacitor losses?" The equation does not seem to take that into account. You would be right - it doesn't. In practice we can ignore the losses in the capacitor because the losses in the inductor swamp the losses in the capacitor. The losses of the capacitor are so small compared to those of the inductor we can ignore the capacitor and just calculate Q using the inductor.

We use tuned circuits and inductors and capacitors to make filters. A filter is a device that responds differently to different frequencies. For example a bandpass filter

passes a narrow band of frequencies in and blocks all others. How "sharp" or "selective" a bandpass filter is at doing this is very much to do with its Q.

Suppose we made a bandpass filter to pass 10.7MHz plus and minus 300kHz. This filter passes a band of frequencies centred on 10.7MHz with a bandwidth of 600kHz. If we did a graph of the filter's frequency response we would get this diagram shown in figure 19-13.

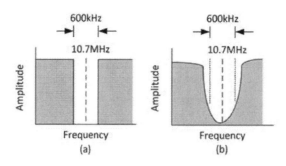

Figure 19-13

In figure 19-13a we have a filter circuit that has no losses or an infinite Q. Notice the sharpness of the filter. It passes exactly 600kHz with no losses and is centred on 10.7MHz. Of course such a filter is not possible. There are always some losses and the Q is not infinite. Figure 19-13b shows a more realistic but nevertheless high Q bandpass filter. We call this sharpness of a filter its "selectivity". The width of the signal that is passed by filter is called its "bandwidth". Bandwidth is related to the Q and frequency of the filter.

$$BW = \frac{f}{Q}$$

There are times when we want to increase the bandwidth of a filter or tuned circuit. We want to make it broader or decrease the selectivity. One of the ways to do this is to reduce the Q. We can reduce the Q by introducing a loss to the circuit in the form of a resistor. Adding a resistor in series or parallel will dissipate heat, increase the losses , lower the Q and broaden the bandwidth.

IMPEDANCE OF TUNED CIRCUITS

When you read 'tuned circuits,' you can also think of resonant circuits as we are discussing one and the same thing. Parallel and series tuned circuits have different properties which allow us to use them in electronics for different purposes.

If there were no losses in tuned circuits (never possible) the following would be true:

Parallel resonant circuits have infinite impedance.
Series resonant circuits have zero impedance.

In the real world though resonant circuits have some losses and so we have to say:

Parallel resonant circuits have a maximum impedance at resonance.
Series resonant circuits have a minimum impedance at resonance.

The impedance type for both series and parallel LC circuits at resonance is resistive. The impedance of a series resonant LC circuit is minimum and resistive. The impedance of a parallel LC circuit at resonance is maximum and resistive.

There is no net reactance at resonance since $X_L=X_C$. That is why we say that resonant circuits are resistive.

IMPEDANCE TYPE ABOVE AND BELOW RESONANCE

First, what do we mean by impedance type?

Impedance is made up of three oppositions:

1. *Resistance (not affected by frequency).*
2. *Inductive reactance - increases with frequency.*
3. *Capacitive reactance - decreases with frequency.*

All of these oppositions are measured in ohms. All oppose current flow. The total combined effect on a circuit of all of these oppositions is called impedance and it too is measured in ohms.

Now think of a circuit at resonance. At resonance $X_L=X_C$. So at resonance the only opposition is resistance, or a better way of putting it, the impedance type of a resonant circuit is resistive. The impedance type of both series and parallel circuits at resonance is resistive because $X_L=X_C$ - Anything which is resonant has no reactance, only resistance.

Impedance type of a series LC circuit - off resonance

A simple way to work out what the impedance type of a series circuit used above or below its resonant frequency, is to think, which reactance (L or C), has the largest voltage drop.

Above resonance X_C will decrease and X_L will increase. Therefore, X_L has the largest voltage drop and the impedance type is inductive. Below resonance the X_C is greater than X_L. The capacitance will have the greater voltage drop and the circuits impedance type is capacitive.

Impedance type of a parallel LC circuit - off resonance

With parallel LC circuits, the largest branch current determines the impedance type. Say to yourself, which branch will have the least reactance and therefore draw the most current. Above resonance X_L increases and X_C decreases and the capacitive branch will draw the most current. Above resonance a parallel circuit is capacitive. Below resonance then, it has to be inductive.

Summary- Impedance Type (i.e. L or C)

Parallel LC circuits:

1. Above their resonant frequency are capacitive.
2. Below their resonant frequency are inductive.

Series LC circuits:

1. Above resonance are inductive.
2. Below resonance are capacitive.

Understanding the impedance type off resonance is very useful. This especially so for series circuits because our antennas are series circuits. If you measure the impedance of your antenna with an impedance meter and it is inductive, then it is too long. If it is capacitive then it is too short.

IMPEDANCE OF LC CIRCUITS VERSUS FREQUENCY

A parallel LC circuit has a maximum impedance at resonance. At resonance, the impedance is maximum and resistive. Either side of resonance the impedance will be less. An LC circuit with no losses has an infinite impedance at resonance. If there was a resistance in parallel with a parallel resonant circuit, then the impedance of the circuit would be equal to the parallel 'R'. For example, an infinite impedance (which is the same as an open circuit) in parallel with a 100kΩ resistor would give a total impedance at resonance of 100kΩ. (Figure 19-14)

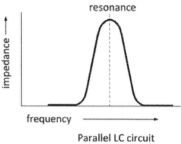

Parallel LC circuit

A series LC circuit has a minimum impedance at resonance. At resonance, the impedance is minimum and resistive. Either side of resonance the impedance will be greater. A series LC circuit with no losses has zero impedance at resonance. If there was a resistance in series with a series resonant circuit, then the impedance of the circuit would be equal to the series 'R'. For example, a zero impedance (which is the same as a short circuit) in series with a 100kΩ resistor would give a total impedance at resonance of 100kΩ. (Figure 19-15)

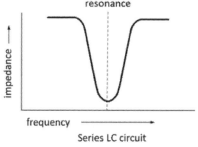

Series LC circuit

THE EFFECT OF LOSSES ON THE 'Q' OF TUNED CIRCUITS

We know that, in the real world, we cannot have parallel or series tuned circuits with no losses at all. The effect of losses on a tuned circuit is to flatten out the impedance curves shown in figure 19-16. The higher the Q the sharper the curve. Tuned circuits can be used to make filters and select radio stations. Sometimes we want the curve (often referred to as the selectivity curve) to be sharp or a high 'Q'. Other times we want it to be flat and not as selective so we may add losses in the form of resistance to flatten the curve out. (Figure 19.16)

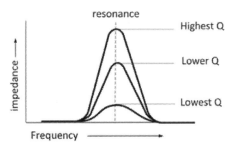

Figure 19-16 shows the curves of a parallel tuned circuit with three values of 'Q'. For a series resonant circuit, the graph would be inverted.

Figure 19-17

What is a flywheel? In mechanics, a flywheel is a wheel with a large mass (weight). If you get the flywheel turning, it will continue to turn for a long time because of its mass. A flywheel is a means of storing energy and releasing it later.

The science fiction writer Isaac Asimov predicted that one day electric power stations would use gigantic flywheels weighing hundreds of thousands of tonnes. Perhaps floating on mercury or using magnetic levitation. Power stations have a demand on the generators that varies enormously. They cannot, particularly with fossil fuel plants, turn the furnaces up and down at a whim. Imagine though if you had a gigantic flywheel, perhaps floating on mercury or on a magnetic field to avoid friction. All of the energy created by the power station is used to make the enormous flywheel move. You would be storing a huge amount of energy in this flywheel. It is so massive that it would continue to rotate for months left untouched. When we wanted energy from the flywheel, we could make contact with it and have it drive a generator for us. So, even though the output of the power station may be irregular, the power available from the flywheel would be regular because of the massive amount of stored energy.

Now a resonant circuit exhibits the same properties of a flywheel. This effect of resonant circuits is called the flywheel effect.

Suppose we want a sinewave to be produced. We know that a resonant circuit will generate a sinewave if it is given energy. We know that the sinewave output from a resonant circuit will dampen if we don't keep on providing it with energy.

Suppose we have pulses of energy, very short pulses and assume we have 1 million of them per second. Let's use these pulses to start a resonant circuit oscillating. Let's make the resonant frequency of the tuned circuit 3 megahertz. This means that the resonant circuit would get a pulse of energy and produce 3 cycles of sinewave output. Now a tuned circuit of reasonable Q will be able to run fine for 3 cycles. Then our next pulse comes along and away the tuned circuit goes again for another three cycles. The output of the resonant circuit is a sinewave at 3MHz. The input is some 'cruddy' pulses at 1MHz.

147

So, the resonant circuit is producing a 'flywheel effect' for us.

There are many situations in electronics where we want to convert pulses to sinewaves or double or triple the frequency of a signal. One way of doing this is the flywheel effect of a resonant circuit.

APPLICATIONS OF RESONANT CIRCUITS

The diagram of figure 19-18 shows an antenna connected to a simple "crystal-set" radio receiver. The first stage of this receiver is L1 and C1 which form a parallel tuned circuit. The parallel tuned circuit is connected between antenna and ground.

Now, remember a parallel tuned circuit has a very high impedance at one frequency only, its resonant frequency.

All of the radio waves passing by the antenna will induce a minuscule voltage into the antenna. Let's say our resonant circuit is tuned to 1000kHz. That is the radio signal that we want to select and reject all others.

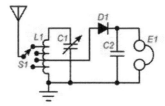

Figure 19-18

A radio wave on 1000kHz passing by the antenna will induce a voltage into it. This voltage will cause a current to flow down the antenna cable and through the tuned circuit to earth. At 1100kHz, the inductor and capacitor are not resonant and the impedance of the parallel circuit will be very low. E=IR or if you like E=IZ (Z=impedance). The voltage created across the LC parallel circuit will be very low as its impedance is low. The same story applies to all other radio signals that induce a voltage into the antenna except at 1000kHz. At 1000kHz, the LC circuit is parallel resonant and will be a very high impedance. The small current through the parallel resonant circuit will produce a significant voltage across it compared to all the other radio signals.

A voltage will appear at the output terminals of the signal that the parallel tuned circuit is tuned to. So here we have the basic method of selecting the desired radio signal from the many that are present at the antenna.

An application for a series resonant circuit:

Recall that at resonance a series tuned circuit has extremely low impedance. If you like, you can think of it as zero impedance or even a short circuit. Zero impedance is like a piece of conductor. If you place a conductor across the back of your TV on the terminals where the signal comes into the set, do you think you would get much of a television picture?

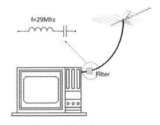

f=29Mhz

Filter

Figure 19-19

Suppose this TV is getting interference from an amateur radio transmission on 29MHz. If we place a series resonant LC circuit across (in shunt) the TV feedline then anything on 29MHz will see a short circuit at the back of the TV and be stopped in its tracks.

The TV signal, which is well above 29MHz will be unaffected. Depending on the Q of the series LC circuit we could expect an attenuation on 29MHz of 100 times or 20dB.

This is an example only and it would work. However when we come to do transmission lines we will find that instead of an LC circuit we can do this with even less effort with a transmission line stub.

A final word on Q and something to remember because it seems to puzzle many; in radio circuits:

Low Q means broad bandwidth or poorer selectivity.
High Q means narrow bandwidth or better selectivity.

Adding losses to any antenna, or tuned circuit causes its Q to decrease. These losses are always through 'resistance' of some sort. The resistance of the wire used in a helically wound vertical, ground losses under an antenna and the resistance of the wire used for an inductor in a tuned circuit are good examples. If the wrong core material is used for an RF inductor or a balun, then the losses can be very high and the Q low. Resistance is sometimes deliberately added to tuned circuits to lower their Q and increase their bandwidth.

DESCRIBING A SELECTIVITY CURVE

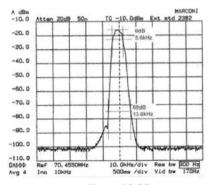

Figure 19-20

It is somewhat difficult to describe an entire selectivity curve using words. One way that is used is to state the curves bandwidth at the 6dB and 60dB points. This method is used to describe the selectivity of a receiver or a filter. Figure 19-20 shows a bandpass on a spectrum analyser. The bandwidth at 6dB is 5.6kHz and at 60dB it is 13.6kHz. This looks like a good double sideband filter or receiver front end selectivity to me.

20 - Power Supplies

THE RECTIFIER DIODE

A rectifier is another name for a diode. I am not going to explain the internal operation of a rectifier now as we will be covering that in full later. I will explain what a rectifier diode does and ask you to accept the internal operation for now.

The schematic symbol of a rectifier is shown in figure 20-1. A rectifier (diode) has two terminals, a cathode and an anode as shown. If a negative voltage is applied to the cathode and a positive to the anode, the diode is said to be forward biased and it will conduct. If it is reverse biased, it will not conduct. So a diode will only conduct current in one direction i.e. from the cathode to anode.

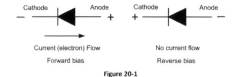

Figure 20-1

PEAK INVERSE VOLTAGE (PIV)

Though we say a diode will not conduct in the reverse direction, there are limits to the reverse electrical pressure that you can apply. Manufacturers of diodes specify a 'peak inverse voltage' (PIV) for each particular diode. The PIV is the maximum reverse bias voltage that the manufacturer guarantees that the diode will withstand. In practice, the PIV is well below the pressure that will cause the diode to breakdown and conduct. However, except in special circumstances, the PIV should never be exceeded.

ELECTRON FLOW VERSUS CONVENTIONAL CURRENT FLOW

Some textbooks show current flow from anode to the cathode. This is called the conventional direction of current flow and is a hangover from the days when the current was thought to flow from positive to negative. Blame Benjamin Franklin, he arbitrarily set the flow of an electric current from positive to negative. That was before J.J. Thompson discovered the electron. Electrical engineers and electricians seem to hang on to this conventional direction of current flow. Electrons flow from negative to positive, from the cathode to the anode. In radio, electronics and communications, 'electron flow' is almost universally used. Electron flow is of course from negative to positive.

The vast majority of 'modern' references agree on the direction of current flow. Even textbooks that use conventional flow acknowledge that this direction (positive to negative) is just a convention. There is no harm in using conventional flow if this is what you prefer. However, for circuit descriptions in this book, electron flow is used.

To help remember which way current flows through a diode look at the symbol - it has an arrow and current flows against the arrow. This will help you not only with the diode but other semiconductor devices to be discussed later.

Let's have a look at the operation of a single diode rectifier if we apply an alternating voltage source to it. Since the diode will only conduct when the potential on the cathode is negative with respect to the anode, only half of the sinewave of alternating voltage will cause current to flow in the circuit of figure 20-2.

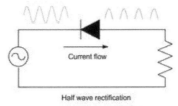

Half wave rectification

Figure 20-2

The current flow in the circuit of figure 20-2 is DC, but it is pulsating DC. The diode refuses to conduct for a complete half cycle of input voltage. The diode conducts (is forward biased) during the half cycle where the cathode is negative and the anode is positive.

This is called halfwave rectification. This terminology rectification has always intrigued me a little. To rectify something means to 'make it right,' to 'fix it up!' This is the origin of this term. The radio pioneers considered converting AC to DC as fixing up the AC voltage, hence the term rectification. I often wonder if Edison had anything to do with the terminology. Edison was strongly opposed to AC power distribution. Tesla and Edison were bitter rivals. It was Tesla who invented and implemented the AC power and distribution system we use today. Edison even electrocuted animals in public performances to demonstrate the dangers of AC. As it turned out, Tesla's AC system was more efficient and therefore adopted. The first power generating plant was constructed at Niagara Falls. Let me make the point that AC does not need 'fixing up', but we are left with this term 'rectification', which I find rather amusing. In many cases, we need to convert DC to AC, so I suppose we should call this un-rectification!

So, in trying to convert AC to DC, all we have managed to do so far is convert AC to half-wave pulsating DC (PDC).

Diodes make it very easy to produce full wave rectification. I am going to start with the most common method, the bridge rectifier. A bridge rectifier has four diodes as shown in figure 20-3. Rather than apply an AC voltage source, I have shown how the bridge network will operate when the two different polarities are applied to it, which in essence is simulating AC.

As you can see from figure 20-3, two of the bridge diodes will always conduct because they have a negative voltage applied to their cathodes with respect to their anodes. Notice how current flows against the arrow. More importantly, notice how current always flows through the load resistor in the same direction, regardless of the polarity applied to the bridge.
We have created full wave rectification.

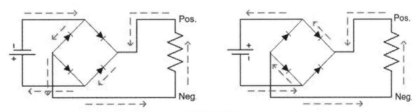

Figure 20-3

You simply must commit the diagram of a bridge rectifier to memory, but be careful, the diodes can all be reversed and you still have a working bridge. I like to think of the four diodes as positions on a clock face. So, we have diodes at 2, 4, 8 and 10 o'clock. The diodes at 2 and 8 o'clock can be reversed, provided the diodes at 10 and 4 o'clock are reversed. You also need to be able to determine the polarity of the output, or the direction of the current through the load. A way to do this easily is to pretend you are an electron out for a stroll. You can go wherever you want provided you go into a cathode and you will seek to get to the positive terminal of the battery or supply. You can then trace out the current flow mentally or on paper and determine if the bridge is drawn correctly.

The bridge I have drawn is a diamond or kite shape, but it can also be drawn as a square. Tracing the current to determine if current will flow through the load in the same direction regardless of polarity, as I have done, is your surest way of determining if the bridge circuit is drawn correctly.

Using a centre tapped transformer for full wave rectification.

A less conventional method (today) of obtaining full wave rectification is to use a centre tapped transformer and only two diodes. I have added some extra detail to the schematic in figure 20-4, which I will fully explain as we go along. A bridge rectifier would not use a centre tapped transformer as shown.

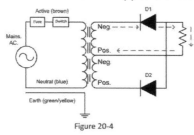

Figure 20-4

I have shown the polarities on the transformer for one-half cycle of AC input from the mains supply. As you can see, the top diode has a negative potential on its cathode so it will conduct with the path of the current as shown. During the next half cycle, all of the polarities will be the opposite of that shown and the bottom diode will conduct. In either case, the current will be the same through the load. So, full wave rectification is achieved again.

Can you see that this circuit is two half wave rectifiers? Each diode rectifies half of the AC sinewave giving full wave rectification. For this reason, the secondary winding has twice the turns compared to a transformer used for a full wave bridge rectifier. Twice as many turns means twice the copper on the secondary, more cost and more weight.

This circuit was once very popular, as it only required two rectifier diodes, and is just as effective as the bridge rectifier. Which one to use comes down to cost and convenience. A centre tapped transformer is more expensive than one without a centre tap. Electron tubes (to be discussed) can also be used as diodes. Electron tubes are relatively expensive compared to semiconductor diodes. So if the power supply was using vacuum tubes, this approach would be the most cost effective, as it requires only two diodes and that outweighs the cost of the centre tapped transformer.

Except in very high voltage applications, the modern trend is to use semiconductor diodes, so this circuit (figure 20-4) is not often used today. Still, you are expected to know it and it does have a useful purpose when using junk box parts. Note also that the total secondary voltage is double that required for the bridge rectifier circuit, as only one half of the secondary is being used on alternate half cycles.

CAPACITIVE FILTERING - WITH A HALF-WAVE RECTIFIER

So far the output of the power supplies has not been smooth DC, which, perhaps with the exception of a simple car battery charger, is our requirement. Adding a capacitor to the output of the power supply will smooth the pulsating DC a great deal. The circuit of figure 20-5 shows half-wave rectified DC fed to a capacitor.

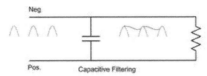

Figure 20-5

If pulsating DC is fed to a capacitor as shown, the capacitor will charge to the peak value of the pulses. Without any load, the capacitor will charge to the peak value and stay there. I have shown the effects of a load. Between the pulses, the capacitor will discharge through the load and the voltage will drop a little. I have left the pulses under the DC output for illustration purposes only. The capacitor is a high value electrolytic - more on this shortly.

The DC output, in this case, is not pure. Pure DC would be a flat line indicating no voltage variation. The DC rises and falls a little, the rate of rise and fall being related to the frequency of the DC pulses. The frequency of the AC from the mains is 50Hz. For every cycle of AC rectified by a half-wave rectifier we get one pulse of DC, so the pulses have a frequency of 50Hz as well. We can, therefore, conclude that the ripple frequency of a half-wave rectifier is the same as the mains frequency i.e. 50Hz.

Just to emphasise that a little, the DC output is not flat, it has a ripple on it. A ripple is a periodic rise and fall of voltage. The rate of rise and fall of the DC output is called the ripple frequency and is equal to the mains frequency, which in Australia is 50Hz.

CAPACITIVE FILTERING - WITH A FULL-WAVE RECTIFIER

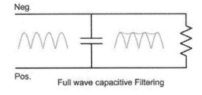

Figure 20-6

The ripple frequency of a full wave rectifier is 100Hz. Full-wave rectification with filtering is much better than a half-wave. However, with capacitive filtering alone there is still some ripple. The frequency of the ripple will always be equal to the mains frequency for half-wave rectification and twice the mains frequency for fullwave rectification.

A COMPLETE CIRCUIT OF A PRACTICAL POWER SUPPLY

Figure 20-7 is the circuit of a full-wave bridge rectifier showing the transformer, the mains connection, a switch, fuse(s) and a capacitor for a filter.

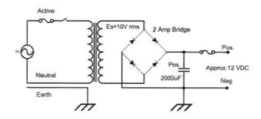

Figure 20-7

Most of this circuit has been explained. I will point out some important features, as such knowledge applies to subjects other than power supplies.

Notice how the secondary voltage Es is shown as 10 volts. This is the RMS value. The output of the power supply is the peak value, RMS x 1.414, or near enough to 12 volts. There will be a small voltage drop across the two conducting diodes. A silicon rectifier diode will have a forward voltage drop of 600-700mV. The negative side (also called the negative rail) of the power supply shows a chassis connection. This just means that if the power supply is in a metal box, the negative terminal is connected to the box. This is not done for safety purposes but for ease of construction, just in the same way that the negative terminal of a car battery is connected to the chassis (body) of the car. Of course, the positive output could be connected to the chassis. This is a construction issue and will depend on the type of semiconductor devices used. However, it is more common for the negative to be connected to the chassis.

I have shown a switch in the active lead of the mains. The switch must always be in the active lead. The earth lead from the mains or power point is connected to the chassis of the power supply. We will have much more to do with safety issues later.

I have shown two fuses. There really only needs to be one. The important thing about fuses is to remember that a fuse operates on current. Should something go wrong in the power supply (or outside the power supply) excessive current will be drawn and the fuse will melt, opening the circuit.

What may not be obvious to you is the difference in choice of fuses - in the primary of the transformer and the output of the power supply. Remember the power in a transformer is the same in the primary and secondary, disregarding losses. However, the voltage and current are not the same. Their product (the power) is the same. Fuses work on current. Say this power supply is 12 volts at 2 amps maximum. The fuse in the secondary circuit will be around 2 amps. It will melt if more than 2 amps are drawn from the supply and stop the supply or the load from being damaged further by excessive current.

Think about the fuse in the primary circuit. It would not be anything like 2 amps. If it was 2 amps, then 480 watts would have to be drawn from the mains before it would melt. I will leave the calculation of the primary fuse to you. As a hint how much power can be drawn from a 12 volt 2 amp power supply?

Let's get a bit more serious about power supply filtering. What you are about to read may go beyond what is required for the examination on power supplies. Though, where concepts are needed to be known, perhaps not in power supplies but in other subjects, I have included the information pertinent to power supplies knowing that this is required knowledge elsewhere.

CAPACITIVE INPUT FILTERING

Capacitance is that property of an AC circuit which opposes **changes of voltage**. If a large value capacitor is connected across the output of the rectifier circuit, i.e. in parallel with the load, the capacitor will smooth out the pulsating DC to DC with a small ripple voltage. The filter capacitor does this by taking energy from the circuit and storing this energy in its electric field. If the voltage across the load tends to fall, the capacitor will discharge some of its energy back into the circuit, which smooths out the voltage. Because a high value of capacitance is required (1000μF to 10,000μF or more), the capacitor has to be an electrolytic type, as this is the only type of capacitor which can provide such a high value of capacitance. Capacitive filtering alone can be used for power supplies which need to deliver only a few amperes of current.

One disadvantage of capacitive input filtering is that of surge current. When the power supply is turned off the capacitor is discharged. When the power supply is first turned on, the capacitor will charge rapidly and draw a large current from the transformer and rectifier circuit. This surge current can be so high that it may exceed the current rating of the rectifiers and damage it, or blow the fuse, whichever comes first. This is not so much an operational problem of the power supply, but a design problem.

To ensure the filter capacitor(s) in a power supply discharges when a power supply is turned off, a high value resistor is often connected across the filter capacitor. A high resistance, say 10kΩ, in parallel with the filter capacitor will have no effect on the operation of the power supply and provide a path for the capacitor to discharge through.

This resistance is known as a bleeder resistor. You may have noticed on some power supplies that have a light to indicate that the supply is turned on, that this light will slowly dim and eventually go out some time after the supply is turned off. This is the filter capacitor discharging. Sometimes the "ON indicator lamp" acts as a bleeder resistor and discharges the filter capacitor.

Best is a high value bleeder resistor connected across the output of the power supply to discharge the filter capacitor(s) when the supply is switched OFF.

INDUCTIVE INPUT FILTERING

Inductance is that property of an AC circuit which opposes **changes of current**. If we are trying to reduce ripple from a power supply, it should be no surprise then, that an inductance can be used to filter the DC output.

When an inductance is connected in series with a rectifier, a filtering or smoothing action results. An inductor used in this manner is always iron core to obtain the high level of inductance required. Inductors used in power supplies for filtering (and some other applications) are referred to as chokes.

Pulses of current through the choke build up a magnetic field around it, taking energy from the circuit to produce the field. As a pulse of current tries to decrease in amplitude, the magnetic field collapses and returns energy to the circuit, thereby tending to hold the current constant. If the inductance is the first component in the circuit after the rectifiers, then this type of filtering circuit is called inductive input filtering. It is not practical to use inductive input filtering alone in a power supply. Inductive filtering is always used in conjunction with capacitive filtering. A disadvantage of inductive input filtering is that there is voltage dropped across the internal resistance of the choke (the resistance of the wire making up the choke). The voltage dropped across the choke reduces the output voltage of the power supply. In practical terms, this means that the secondary voltage of the transformer has to be a little higher in voltage output to compensate.

SWINGING CHOKES

You will recall from our study of inductance that the purpose of the iron core is to increase the value of the inductance. The magnetic iron core does this by concentrating the magnetic lines of force. Usually, the magnetic iron core of a choke has no air gap to prevent the core from magnetically saturating. In the swinging choke, the magnetic iron core has little or no air gap. This means that the magnetic iron core will begin to saturate at average current values for the power supply. When low current flows through a swinging choke, it has a high inductance and filters effectively. With high current the magnetic core saturates and has less choking effect. Thus, with a light load and little current, the swinging choke has a high reactance and develops a significant voltage drop across it. When the load increases, the choke saturates and has less reactance and consequently less voltage drop across it. This means the voltage output from the power supply tends to remain constant under varying loads, improving its regulation. Typically, a swinging choke will change (swing) from 5H for a heavy load (high current) to 20H for a light load.

COMBINATIONS OF CAPACITANCE AND INDUCTANCE

For better regulation, combinations of chokes and filter capacitors are used. It is not necessary for you to remember these combinations (for exam purposes). It is important to bear in mind that whatever filter type is used, it will get its name from the first component seen, looking from the rectifier towards the filter.

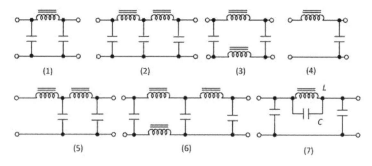

Figure 20-8

If the first component is an inductor(s), then it is an inductive input filter. Likewise, if the first component is a capacitor, it is called a capacitive input filter. You may be asked the advantages and disadvantages of each type.

(1) Pi - Capacitive input filter.
(2) Capacitive input filter.
(3) Capacitive input filter.
(4) Inductive input filter.
(5) Inductive input filter.
(6) Capacitive input filter.
(7) Capacitive input filter.

All of the power supply filter circuits shown in figure 20-8 use the principles we have discussed (except for circuit 20-8(7)). The purpose of a power supply is to provide smooth DC output, that is, a constant voltage with varying load conditions. In all circuits, each parallel capacitor has a capacitance that opposes change in voltage and the inductance of the inductors oppose change in current.

156

Figure 20-8(7) uses another principle, which you don't often see in power supplies. However I will discuss it as the principle is one we need to know for other purposes. In circuit 20-8(7) the parallel inductor and capacitor are resonant at the ripple frequency. A parallel tuned circuit has very high impedance only at its resonant frequency. The parallel LC is resonant at the ripple frequency. Any 100Hz ripple (assuming full-wave rectification) will be blocked significantly from passing through this LC combination. An ingenious technique.

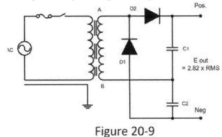

Figure 20-9

THE VOLTAGE DOUBLER (FULLWAVE)

The diagram in figure 20-9 is that of a full-wave voltage doubler. With a standard fullwave power supply (e.g. bridge) the output voltage is approximately the peak value of the RMS secondary voltage, or RMS x 1.414.

When lines are crossed without a dot as shown, it means the wires are not connected. When the AC voltage makes point 'A' negative and 'B' positive, D1 will conduct and charge C2. When the AC voltage makes 'B' negative and 'A' positive, then D2 will conduct and charge C1. The capacitors C1 and C2 are in series and so the voltage at the output is the sum of the voltage on C1 and C2, or 2.82 times the RMS secondary voltage.

The voltage doubler can't deliver very high currents. However, it is an excellent method of obtaining the higher voltages required for an electron tube.

EQUALISING RESISTORS AND TRANSIENT PROTECTION CAPACITORS

In very high voltage power supply circuits, many rectifier diodes are often connected in series to increase the peak inverse voltage (PIV) rating. If a rectifier has a PIV of 400 volts, then 'theoretically' two such rectifiers connected in series should have a PIV of 800 volts. More diodes can be added in series for even higher PIV. However, for proper series rectifier operation it is important that the PIV be divided equally amongst the individual diodes. If it is not done, one or more diodes in the string may be subjected to a greater PIV than its maximum rating and as a result may be destroyed. As most failures of this type lead to the diode junction going short circuit, the PIV of the remaining diodes in the string is raised, making each diode subject to a greater value of PIV. Failure of a single diode in the series (stack) can, therefore, lead to a 'domino effect', which will destroy the remaining diodes.

Forced voltage distribution in the stack is necessary when the diodes vary appreciably in their reverse resistance.

The resistors in figure 20-10 provide PIV equalisation across the two diodes.

The reverse resistance of the diodes will be very high, in the order of megohms. Now, 470kΩ in parallel with megohms is roughly still 470kΩ. So irrespective of the diode's reverse resistance the PIV is distributed roughly equally across each diode.

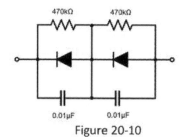

Figure 20-10

TRANSIENT PROTECTION CAPACITORS

Power supplies, high voltage ones in particular, are prone to producing high voltage spikes called transients. These transients are caused primarily because of inductive effects of the circuit (the secondary of the transformer). These are very short duration high voltage transients which can destroy the rectifier diodes. The capacitors shown across the diode in figure 20-10 above will absorb or smooth out any transients. Recall that capacitance is that property that opposes voltage change.

In summary, equalising resistors distribute the applied voltage almost equally across a string of diodes when they are reverse biased. This ensures that each diode, irrespective of its reverse resistance, will share the same amount of reverse voltage drop. Transient protection capacitors absorb short duration high voltage transients caused by inductive effects in the circuit.

ELECTRONIC POWER SUPPLY REGULATION

For power supplies that require excellent regulation, electronic regulation can be added to the filter network. This consists of either an electronic circuit comprised of discrete components or, far more commonly these days, an integrated circuit that contains all the necessary electronics.

There are three basic types of regulation:

Series Regulation - where the regulator is in series with the current path.
Shunt Regulation - where the regulator is placed in parallel or 'shunted' across the power supply output.
Switching regulators -these are the most efficient - that is they dissipate the least amount of wasted power.

These days series regulation is almost always used. The disadvantage with the shunt regulator is that the power dissipated by the regulator circuit is much higher. This translates to lower power supply efficiency and more heat to get rid of in the regulator circuit.

Figure 20-11 is a schematic diagram of a power supply that uses a three-terminal integrated voltage regulator; this is a series regulator. More efficient than a shunt regulator.

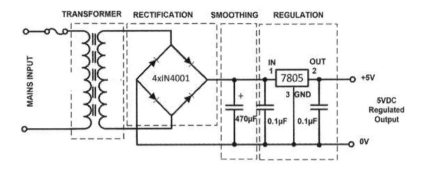

Figure 20-11

The 7805 voltage regulator is just one of many off the shelf integrated voltage regulators.

The IC 7805 is a 5V voltage regulator that regulates the voltage output to 5V. It comes with a provision to add a heatsink. The maximum value for input to the voltage regulator is 35V. It can provide a constant steady voltage level of 5V for higher voltage input to the threshold limit of 35V. If the voltage is near to 7.5V, then it does not produce any heat and hence no need for a heatsink. If the voltage is too high, then excessive power is dissipated as heat from the 7805.

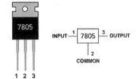

Figure 20-12

IC 7805 is a series of 78XX voltage regulators. It's a standard, from the name the last two digits '05' denotes the amount of voltage that it regulates. Hence, a 7805 would regulate 5v and 7806 would regulate 6V and so on.

The shunt regulator method shown in figure 20-13 is only used for very low power circuits as shunt regulation is the least efficient of all types of regulation.

A good practical example of a shunt regulator and the only one you need to know is that of a zener diode along with its current limiting resistance.

We have not discussed the operation of a zener diode yet - we will do so in the semiconductors chapter. For now, a zener diode is one that is meant to be operated with reverse bias and beyond its PIV or breakdown voltage. Usually, a diode would be destroyed if operated beyond breakdown. The series resistance R1 (refer to figure 20- 13) limits the reverse current so that the zener diode is not destroyed.

Under such conditions, the zener diode will have a constant voltage drop across it. Zeners come in a range of voltages sometimes a bit odd like 9.1 volts. Zener regulators are used a lot inside transceivers to regulate the voltage to the active devices. A typical example is an oscillator which is low power consumption and must have a stable voltage.

159

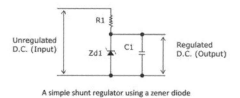

A simple shunt regulator using a zener diode

Figure 20-13

Assuming this zener is 9.1V then as the unregulated input voltage varies from say 10 to 15 volts, the voltage across the zener will remain (almost) constant at 9.1 volts. However, the voltage across R1 will vary. The sum of the zener voltage (9.1 V) and the voltage across R1 will equal the unregulated input voltage as this is a series circuit.

The zener circuit draws maximum current when there is no load as shown now in figure 20-13. This is why shunt regulators are so inefficient. They are dissipating maximum power when doing no work. When a load is applied some of the current going through the zener will now go through the load reducing the heat dissipated by the zener. The sum of the currents through the load and the zener is constant. A zener shunt regulator circuit can be converted into a series regulator circuit by the addition of pass transistors.

A CONSTANT CURRENT SOURCE

This simple circuit (figure 20-14) is a favourite of mine, as I have found so many uses for it. First, allow me to explain what constant current means. The purpose of most power supplies is to maintain a constant regulated DC voltage output. For example, a radio transceiver may require around 13.8 volts DC. The transceiver, when in use, will

draw whatever current it requires from the power supply, so the current varies. The purpose of the power supply is to maintain the constant voltage output of 13.8 volts, regardless of the changing current demands of the transceiver.

Now, there are situations where you don't care about the voltage output of a power supply. What you care about is current. This may seem a little strange because usually, we apply a voltage to a load and the load determines the current. Suppose we want to force a particular current through the load irrespective of what the load wants.

A classic example is charging slow charging lithium batteries. These are very common in portable radio communications equipment. Lithium cells can be slow charged at one tenth of their amp-hour rating for 14 hours (if fully discharged). So a 500mA hour cell or battery should be charged at 500/10 = 50mA for 14 hours. This cannot be done with a constant voltage output power supply. We need a constant current supply. A three terminal regulator can be used to do this for a cost of less than $2. Commercial devices for charging lithium cells (including cell phone batteries) are mostly constant current chargers. You can charge Lithium cells at a much higher rate but in this case the temperature and cell voltage should be monitored.

The resistor is in series with the load and the three terminal regulator is producing a constant voltage across it. The resistor will, therefore, have a constant current through it. Since it is in series with the load, the load will have the same current as the resistor.

All you need to do is buy a regulator, which will handle the current you want and calculate the value of the resistor and its minimum wattage.

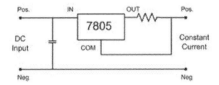

Figure 20-14

POWER SUPPLIES - PASS TRANSISTORS

The zener diode cannot handle a significant amount of current, usually no more than a few hundred milliamps. Also, a zener diode is not a perfect regulator. There is a way of improving the current handling capabilities of a zener regulator circuit and improve regulation at the same time. This is done by adding a pass transistor.

Figure 20-15 is the schematic diagram of a zener regulator with a pass transistor added. You will recognise Rs as the usual current limiting resistance that prevents the zener from being destroyed. The zener is reverse biased, with a positive voltage applied to its cathode.

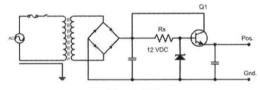

Figure 20-15

The zener regulated voltage drives the base of a transistor connected as an emitter follower (common collector). The output of the emitter follower goes to the load. Because of the current gain (Greek letter beta; β) of the emitter follower, any change in load current is reduced by a factor of β, This means voltage regulation is improved by a factor of β and that is a significant improvement.

Use of the pass transistor increases the maximum load current the regulator can handle. You need to be able to identify this circuit as a zener regulator with a pass transistor. The purpose again - to improve regulation and the current handling capabilities of the zener.

A pass transistor can be added to a three terminal voltage regulator in much the same way and for the same reasons. We will be discussing β and transistors more deeply in the chapter on semiconductors and amplifiers.

THE DARLINGTON PAIR

Two transistors can be connected in a configuration known as a Darlington Pair. As far as I am aware, you do not even have to know what a darlington pair is. We are jumping ahead of ourselves, but it is best, like the series pass transistor to mention them now. The darlington pair is an improvement on the single pass transistor.

Figure 20-16

The higher the β; the higher the input impedance of an emitter follower. One way to increase current gain or β is with a darlington pair, two transistors of the one type connected as shown in figure 20-16.

The overall current gain is equal to the product of the two individual beta. Transistor manufacturers can put a darlington pair inside a single transistor package.

Notice how the external connections to the two transistors are a base, emitter and collector.

A traditional application of a darlington pair is a 'pass transistor' in a power supply. A darlington pair can also be created with two PNP transistors.

A POWER SUPPLY WITH PASS TRANSISTORS (Darlington Pair)

All of the regulator circuits we have looked at are series regulators. The pass transistors are in series with the load. Series regulators are more efficient than the shunt type. The shunt type regulator (which is in shunt or parallel with the load) is rarely used, except in low power circuits, as they dissipate too much power.

Figure 20-17 is a full-wave power supply. It uses two diodes. Therefore, it must have a centre tapped transformer. C1 and C2 are both filter capacitors, probably around 1000 microfarads and therefore electrolytic. R1 is the current limiting resistor for the zener regulator D3.

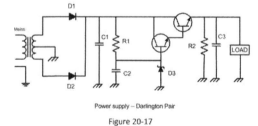

Power supply – Darlington Pair

Figure 20-17

The zener regulator has a darlington pair as the pass transistors. The darlington pair does a superior job compared to a single transistor, improving the regulation and the current handling abilities of this power supply significantly. Read the section on pass transistors and the darlington pair again if you need to.

R2 supplies a light load on the power supply should the 'real' load not be connected. Its value would be about 1kΩ. C3 is an RF bypass capacitor about 0.01μF.

Is the zener diode reverse biased? Yes, the positive rail (line) is at the top of the circuit. See if you can deduce this from the power supply diodes and how they would conduct and charge C1. R1 is a current limiting resistor for the zener diode.

SWITCHING REGULATORS (BUCK CONVERTERS)

As the name infers switching regulators regulate voltage by some form of switching. The best way to understanding this type of regulator is to start from the beginning and develop the principles involved. The type of regulator we will develop is the **buck converter**. This type of converter is the most commonly used in radio communications as it converts a high voltage to a low voltage. One way we could control the voltage to a load could be with a resistor as shown in figure 20-18.

Figure 20-18

The problem with this method of voltage regulation is very poor efficiency. Resistors dissipate energy in the form of heat and this energy is lost. A controlled variable resistance is essentially how series linear regulators work. With switching regulators we are endeavouring to take as much resistance out of the regulator circuit as possible between the source and the load.

Figure 20-19

When the switch in figure 20-19 is open the load receives 0V. When the switch is closed the load receives 10V. The switch can be operated continuously and quickly if we wanted at high frequency. What would be the average voltage across the load? The average load voltage would be 5V. However this is not a good voltage output as there are deviations in voltage between 0 and 10V. We have achieved something. The switch does not dissipate any power as the resistor did in figure 20-18. Perhaps we could smooth the voltage out across the load by adding a capacitor as in figure 20-20.

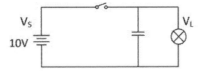

Figure 20-20

Figure 20-20 certainly looks like a step forward. Capacitance is the electric property that opposes **changes of voltage**. Since the capacitor is connected across the load it should oppose changes to load voltage. We still have not used resistance so our circuit efficiency is high. The problem with this circuit is when the switch is turned ON there is a rush of high current into the capacitor. The time constant of the capacitor is short and we do not want to lengthen it by added resistance that we are trying to avoid. The capacitor is smoothing out the load voltage but we need to stop the excessive surge current when the switch is closed.

We could do this by adding a series inductor as shown in figure 20-21. Inductance is the electric property that opposes **changes of current**. The inductor will then control the surge current by storing energy in its magnetic field.

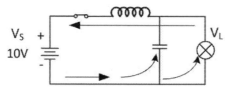

Figure 20-21

Now when the switch is closed (figure 20-21) current flows as shown. There is no capacitive surge current as the inductor opposes this rapid change of current. The load is supplied power from the source. The capacitor is charged and stores energy. The magnetic field about the inductor expands and stores energy. All is looking better. What happens when we open the switch?

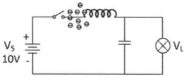

Figure 20-22

Inductance is the electric property that opposes changes of current. We have just opened the switch and demanded that the current stop. The inductor will create a back emf that will oppose our attempt to change current. The polarity of that back emf will be to keep the current going in the direction it was before the switch was opened. A very large voltage will be developed across the inductor. How large? As large as it takes to get rid of the stored energy. The switch will most likely arc across. If a semiconductor switch is used it could be destroyed. The solution is to give the energy of the inductor somewhere to go.

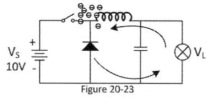

Figure 20-23

With a diode connected as in figure 20-23 there is now a discharge path for both the inductor and the capacitor through the diode and the load when the switch is in the OFF position. We need to look at what the diode does when the switch is in the ON position.

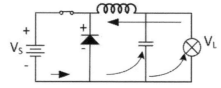

Figure 20-24

When the switch is closed the diode is reverse biased. The diode is then an open switch and is essentially out of circuit.

Recapping, during the ON phases the load, capacitor and inductor are supplied energy from the source and the diode is out of circuit. During the OFF phases the diode is forward biased and conducts now providing a path for the inductor and capacitor to dump their energy by allowing a current path through the load. We now have all the essential topology of a buck converter.

The output voltage depends on the percentage of time the switch is ON. In a practical buck converter we would use a BJT transistor or a MOSFET as the switch. We control the ON/OFF ratio of the switch by a pulse width modulation (PWM) on the control of the switch (gate or base). The switch will operate at anywhere from 60 to 200kHz. The diode would be a schottky type for fast switch and low forward voltage drop.

We would also want to monitor or sense the output voltage and have that control the ON/OFF switching ratio of the semiconductor switch.

It would not be difficult to build a buck converter. We do not have to go to component level as there are many off the shelf switch mode converter integrated circuits and they are inexpensive.

DIAGRAM - BUCK CONVERTER

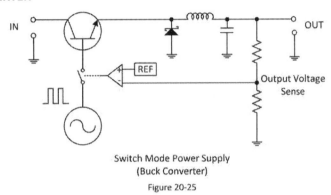

Switch Mode Power Supply
(Buck Converter)
Figure 20-25

The only difference here compared to what we have just discussed is that now we have a voltage divider on the output which is fed back to an operational amplifier and a reference voltage. The oscillator provides the switching square wave and the operational amplifier turns on the square in bursts to the NPN transistor. This transistor would be a darlington pair - see figure 20-16. The diode is a schottky.

A Switch Mode Power Supply (SMPS) that converts to a lower voltage than the input is a buck converter as discussed here. An SMPS using a boost converter outputs a higher voltage than the input. Boost converters operate in a similar way to buck converters. They use the same components organised in a slightly different way. Many off the shelf integrated circuits can be operated in buck or boost mode.

SMPS are more efficient than their linear counterparts. Efficiencies exceeding 90% are possible. SMPS are smaller and much lighter than linear power supplies. They are also less expensive to build. The cost and weight saving is primarily due to the SMPS not requiring a mains transformer and large heat-sinks.

The mains is usually connected directly to the SMPS where there is a bridge rectifier but no transformer. This means the danger of electric shock is higher. Unless you are experienced in working on high voltages an SMPS should always have the power removed when fault finding.

THE MAINS SUPPLY

We have covered in earlier chapters the frequency, voltage relationships, RMS, peak, and the wave shape of the mains. What we have not covered is the wiring of the mains.

Household wiring is a parallel connection. If it were not, when we switch an extra light on, all other lights would dim a bit (series circuit).

The three wires in the (Australian) mains circuit are:

The Active - Colour **Brown** (old red).
The Neutral - Colour **Blue** (old black).
The Earth - Colour Green/Yellow (old green).

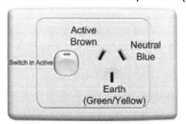

Figure 20-26

These names are a little confusing. First of all, active and neutral deliver the 240 volts AC throughout your home. There are 240 volts AC RMS between active and neutral. The standard for the mains voltage in Australia is really 230V AC RMS but this varies a lot with load and still referred to as the 240V AC Mains.

The earth is not ground in the sense that radio operators call ground. Earth is just that, EARTH, dirt if you like. The earth connection is usually somewhere near your power box and is a copper rod going into the ground.

Also, one side of the mains is connected to the earth as well. This earthed side of the mains is called neutral. The neutral is connected to the 'neutral bar' in the switchboard and the neutral bar to the electrical ground. This means that there must be 240 volts AC RMS between active and earth. The potential difference between neutral and earth should be zero volts, but they may be a few volts (up to 5-6) due to capacitive coupling.

The locations of these wires are shown for a power point (socket) in figure 20-26. The active is on the left on a power point (socket). The active is on the right looking at a plug and on the left looking at the wall socket.

The photo in figure 20-27 shows the connections for both a plug and a socket (extension lead).

Should the active wire come adrift and contact the metal chassis of any appliance a fuse will blow. The metal (conductive) parts of appliances are connected to earth, as is the neutral. Should an active touch a metal part and therefore potentially 'you', the active and neutral are effectively short circuited and a fuse will blow.

All being well since there is no potential difference between neutral and earth you should be able to touch the neutral wire in your house and nothing will happen - never put this to the test, though by all means, use a voltmeter to check.

Figure 20-27

Every power switch in your house should be in the active wire.

It is extremely dangerous to assume that the wiring in your home is correct. If someone has wired a switch into the neutral wire instead of the active, then the appliance will still work. However, the active will still be going to it when it is switched off. For example, if someone has placed the switch in the neutral wire of a light, the light will still work as it should (as will all other appliances). If you turn the light switch OFF and it is wired incorrectly like this, the active is still present at the light socket. If you touch the active terminal in the light socket, you will receive an electric shock that could be fatal.

How easy it is for someone to reverse the active and neutral accidentally or carelessly on the plug or socket of an extension lead. Doing so could be fatal. In passing, electrocution is death by electric shock. You cannot be electrocuted and survive.

You need to know the potential difference between each wire in your household mains. This is not so you can work on your house mains. It is necessary for you to know this if you build equipment that you connect to the mains. You need to know how to test a power point with a voltmeter. This is safe provided you do not contact the metal parts of your multimeter leads. If you measure less than 240V between active to earth and active to neutral, it may mean you have a problem. Any voltage above a few volts between earth and neutral indicates a problem.

If you are in a new home, you will have an **earth leakage detector** fitted. Older homes may not have them, so it is your choice to have one fitted by an electrician.

Only qualified electricians should work on the household wiring. Being a electronics hobbyist or a professor of physics does not qualify you to tamper with house electrical wiring. This part of the chapter is not a definitive explanation of safety procedures. The best procedure is to remove equipment from the mains before working on it. . Beware when working on unknown equipment which is to be connected to the mains. Perform an electrical safety check if you are competent to do so. If you are not competent then have someone who is test the equipment for you.

Different countries will have specific Federal, State, or Local government laws. You need to know your local requirements regarding electrical safety.

21 – Semiconductors I

The objective of this chapter is to cover how semiconductors work. Discuss the operation of bipolar junction transistors (BJTs), FETs MOSFETs and SCRs, as well as a few of the characteristic circuits in which they are employed.

Solid state devices

Solid state devices are much smaller physically, more rugged and lighter than vacuum types, but cannot withstand heat as well, changing their characteristics as they warm. However, many newer special solid state devices will operate at much higher frequencies than any vacuum device.

SEMICONDUCTORS

The outer orbit, or valence electrons of some atoms, such as metals (conductors), can be detached with relative ease at almost any temperature and may be called free electrons. This is why metals are good conductors. These valence electrons can move outward from a standard outer orbit level into a conduction level or band, from which they can be easily dislodged. Such materials make good electrical conductors. Other substances such as glass, rubber and plastics, have no free valence electrons in their outer conduction bands at room temperatures and are therefore good insulators.

A few materials have a limited number of electrons (4) in the conduction level (outer orbit or valence band) at room temperatures and are called semiconductors. Applying energy in the form of photons (small packets of light or heat energy) to the valence electrons moves some of them up into the conduction band and the semiconductors then become better electrical conductors. Energy of some form is required to raise semiconductor electrons to a conduction level. Conversely, if an electron drops to the valence level from a higher conduction level, it will radiate energy in some high frequency form such as heat, light, infrared, ultraviolet and, if the fall is significant enough, X-rays.

The semiconductors, germanium and silicon have four outer valence electrons. Semiconductors can be laboratory grown, this makes mass production of them easy.

COVALENT BONDING

In nature atoms are more stable and 'like' to have eight valence (outer) electrons called an octet. Chlorine has seven valence electrons. If two Chlorine atoms are brought near each other, they will snap together and form a Chlorine molecule. Each atom is sharing one of its valence electrons with the other. This covalent bonding makes the Chlorine molecule more stable.

Noble gases have eight valence electrons called an octet, this is what makes noble gases so stable.

In the case of germanium and silicon which only have four valence electrons. A special arrangement occurs when they rub shoulders. Germanium or silicon in pure form creates what is called a crystal lattice structure.

The four valence electrons of each atom in a crystal lattice structure share themselves with the adjacent orbits of all other electrons in the crystal lattice structure. In this way by borrowing electrons from neighbouring atoms a type of shared valency of eight is created. Consequently, the valence electron arrangement is very stable and it is difficult to make these electrons participate in current flow. The great breakthrough in physics (electronics) was the discovery of how the characteristics of pure germanium or silicon could be changed dramatically by adding impurities (other atoms) to the crystal. Adding impurities disturbs or upsets the crystal lattice structure.

INTRINSIC SEMICONDUCTOR LATTICE

A perfectly formed intrinsic semiconductor crystal lattice is illustrated in figure 24-1. Such a crystal acts more like an insulator than a conductor at room temperature. I have drawn the crystal in only two dimensions (not being much of an artist). Shown are the valence band (the outer orbit of each atom) and the four valence electrons. The valence electrons are of course not stationary, but orbiting around the atom as if on the surface of a sphere. The sharing of valence electrons is called covalent bonding. This arrangement is very stable electrically.

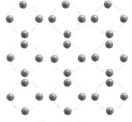

Figure 21-1

Electrons are locked into the crystal lattice and at normal temperatures the crystal is neither a good insulator nor conductor.

DOPING (N-type)

Doping is the process of deliberately adding impurities to the crystal during manufacture. Say we added some arsenic, which has five valence electrons to a pure semiconductor crystal. The ratio of germanium to arsenic is about a million to one.

Arsenic is pentavalent (five outer electrons) and cannot fit into the crystal lattice structure. What happens is four of the arsenic electrons participate in the sharing (covalent bonding) and one is left out!

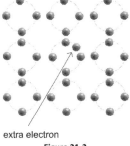

extra electron
Figure 21-2

The crystal lattice shown in figure 24-2 has one atom in a million with an excess outer ring electron not being tightly held. This type of doped semiconductor is called N-type.

It should be understood that 'N1 type semiconductor does not have extra electrons electrically. N-type semiconductor does not have a negative charge. What is 'extra' in N-type material is electrons which do not fit into the crystal lattice structure. These extra electrons are not locked into the crystal lattice structure, so they are much easier to move.

When an electrostatic field (by application of an emf) is developed across such arsenic-doped germanium, a current will flow. The N-type semiconductor is about 1,000 times better as a conductor than the intrinsic semiconductor. Doped germanium with such relatively free electrons is known as N germanium and is a reasonably good conductor. To form N-silicon, phosphorus can be used as the dopant - you do not need to remember the dopant. Merely by adding a small amount of impure pentavalent atoms to a pure crystal, we convert it into a conductor by disrupting the harmony of the crystal lattice structure.

DOPING (P-Type)

When germanium is doped with gallium, which has three valence electrons (tri-valent), the crystal lattice is again disrupted. This time, there is an area, or hole, in the crystal lattice structure, a region that apparently lacks an electron. While the hole may not actually be positive, at least it is an area in which electrons might be repelled to by a negative charge. This positive appearing semiconductor material is called P-germanium or just P-type.

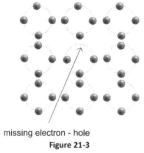

missing electron - hole
Figure 21-3

When an electrostatic field is impressed across a P-type semiconductor, the hole areas act as stepping stones for electron travel through the material. It can be said that hole current flows in a direction opposite to the electron flow. Note that both N-germanium and P-germanium have zero electric charge because both have an equal number of electrons and protons in all of their atoms. One dopant used to produce P silicon is boron.

HOLES

I am going to talk about holes for a bit, as it seems to be a stumbling block for many. I have drawn the hole as a gap in figure 24-3. The hole is a missing electron in the crystal lattice structure which destroys the crystals insulating properties. It is very easy with N-type material to visualise that electron flow can take place. With a hole, it is a little harder and I find some textbooks a little confusing on this issue. A hole is a hole in the crystal lattice structure. Because the lattice is not complete in the location where a hole is, electrons can move into the hole and in doing so, they create a hole from whence they came.

SOME FOOD FOR THOUGHT

A hole is a missing electron in the crystal lattice structure. A hole is not a positive charge.
A hole like any hole can be filled.

A hole can be filled with an electron.

When an electron does fill a hole - then the filled hole disappears, but where the electron came from there is now a hole.

A hole can be thought of as positive for behavioural description purposes.

An electric current is an ordered movement of electrons.

Holes allow electrons to move in the crystal by giving them somewhere to go i.e. filling a hole.

When an electron moves to fill a hole it leaves a hole from whence it came.

Figure 21-4

I think most of us have seen the toy shown in figure 24-4. A flat panel of plastic squares with pictures or numbers on them and the objective is to move the squares around into some order, either to get the numbers in order or to make a picture. Would you be able to slide the squares around if the game was made without a missing square? No, of course not. By leaving a square out, leaving a hole in the puzzle, it makes it possible to slide the squares around by moving them into a hole. In moving a square into a hole you create a hole, making it possible to slide other squares into that hole. Think of the squares as electrons and their ability to move is made possible by the presence of the hole (missing square).

Slide 14 (figure 24-4) could be moved straight up. That would leave a hole where slide 14 was. Now, if you were to put this puzzle on autopilot and sit back and watch it, what would you notice about the way the hole moves? It moves in the opposite direction to the slides. So if the numbered squares are electrons the hole is behaving like a positive charge in that it moves in the opposite direction to electrons. Some references do not explain this very well and go on to talk about "hole current" moving from positive to negative, I believe this to be confusing. Electrons are the only current. Though I will be talking shortly about holes moving, holes do really move when you fill them. However, all of the real moving is done by electrons falling into holes in a P-type semiconductor.

The P-type semiconductor material is a conductor because of the presence of holes in

the crystal lattice structure. Doped silicon has considerably more resistance than germanium and finds its uses in higher voltage and current applications. Also, it does not change its resistance when heated and it can withstand greater temperatures without its crystalline structure being destroyed.

So what?

Well, we have not seemed to have achieved much have we? We have taken a perfectly good semi-insulator (semiconductor) and turned it into a conductor by adding pentavalent impurities (N-type) or trivalent impurities (P-type). The magic starts when we combine the two, that is, we make one side of the crystal N-type and the other P-type. They are not made separately and then stuck together; crystals are grown and doped on different sides of the same crystal in the laboratory.

SOLID-STATE DIODES

Before we start, the term solid state is only used because the alternative devices before them were the electron tube devices.

Let's take a piece of N type and P type semiconductors and join them together. The area in which the N and P substances join is called the junction. We don't really take one of each and physically put them together. It is achieved by doping different sections of the same piece of intrinsic semiconductor.

Some of the relatively free electrons in the N-type material at the junction fall into some of the holes in the P-type material at the junction. So, right at the junction, there are no free electrons (in the crystal lattice structure) and no holes, as free electrons from the N material have filled some of the holes in the P material. This creates a region at the junction, which has neither free electrons nor holes and it is called the *depletion zone*.

This develops an area at the junction that is actually slightly negative on the P side (because electrons have filled holes) of the junction and slightly positive on the N side (because electrons have left to go and fill holes). This produces a barrier to any further electron flow of about 0.2 V with germanium and 0.6-0.7 V with silicon diodes. This small potential is called the *barrier potential*.

We have created a semiconductor diode (PN junction). In figure 24-5 the depletion zone is shown at the junction drawn as a small band.

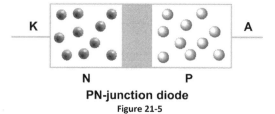

PN-junction diode
Figure 21-5

The whole (no pun intended) diagram is exaggerated greatly as the depletion zone is extremely narrow and there are many more holes and electrons in a real PN junction. Don't forget the diode in figure 24-5 right now has a very small potential across it called the barrier potential. We can't do any work with that voltage. It is not a small battery; it is a weak electrical pressure, created by hole and electrons recombining at the junction. However, this barrier voltage is real and affects how the PN junction diode operates.

Applying a voltage to a PN-Junction diode

The diode of figure 24-6 has the polarity applied to it as shown, The N; end of the diode is called the cathode (K) and the P; the end is called the anode. Remembering electron tubes, if we applied negative to an anode and positive to a cathode would we get conduction? No, we would not. This is called reverse bias.

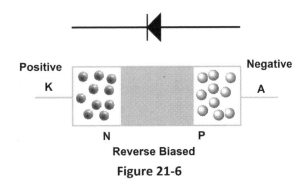

Reverse Biased

Figure 21-6

REVERSE BIAS

In figure 24-6, we have applied an external electrical pressure. Electrons are attracted toward the positive terminal whilst holes, thinking of them as if they were positive charges, are attracted to the negative terminal.

No current flows through the PN junction, which from now on, we will call a diode. The depletion zone is widened as electrons and holes are moving away from the junction. The diode is said to be reverse biased. I have drawn the schematic symbol of a diode, so you can see that for reverse bias, positive is connected to the cathode and negative to the anode. The left-hand side of the diode shown is the cathode.

Recall when we discussed power supplies without really describing how a diode worked. I suggested you look at the diode schematic symbol as an arrow. In figure 24- 6, the arrow is pointing to the left. I asked you to remember that conduction was only possible in the opposite direction of the arrow. Electrons can only flow in the direction cathode-to-anode, against the arrow.

We have said that no current flows through the diode when it is reverse biased. There is an extremely low leakage current which for most practical purposes can be considered to be zero. Also, if you increase the reverse bias voltage high enough you will force breakdown conduction an damage the diode. This reverse breakdown voltage is called peak inverse voltage (PIV).

The reason why you do get a reverse leakage current in a diode is that some of the N material will have just a few holes in it and some of the P material will have some electrons in it. These are due to extremely small amounts of contaminants. So there is a minor 'ghost' diode the opposite way around to the 'real' diode. These contaminants are called minority current carriers. The real holes and electrons are called the majority current carriers. If you like, you have a majority diode one way and a minority diode the other way. When the majority diode is reversed biased the minority diode is forward biased and this accounts for the small leakage current.

The reverse leakage current of a PN junction increases with temperature. Reverse biased PN junctions can be used to measure temperature, by amplifying and measuring the reverse leakage current.

FORWARD BIAS

Now let's reverse the polarity. I know you are aware that the diode is going to conduct but let's look at exactly what goes on. We are going to think of the holes as positive charges when we really know they aren't - we covered that issue with the 'puzzle' example earlier.

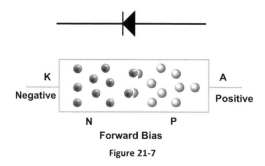

Forward Bias

Figure 21-7

We have now applied a negative potential to the N material (cathode) and positive to the P material (anode). Provided this potential is greater than the barrier potential (0.2V germanium, 0.6V silicon) the depletion zone will be flooded with electrons and the diode will conduct as shown in figure 24-7.

You may well ask, "why don't all the electrons move across the junction and fill up all the holes." We all know that current flow is electrons. In figure 24-7 the right-hand side is P-type (holes) and electrons leave the P-type anode and flow to the positive terminal of the battery. Every electron that leaves the anode creates a hole. The holes move toward the junction to be filled by more electrons.

Such a diode could be used in a rectifier circuit. All diodes have a maximum current rating as well as a peak inverse voltage rating (PIV). A diode, for example, may be rated at 1 amp 400V PIV. Diodes come in many shapes, sizes and package types. All of them more or less work on the principles we have described here. Even LED's (light emitting diodes) are just special PN junction diodes.

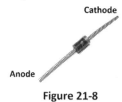

Figure 21-8

Figure 24-8 is a typical rectifier diode that could be used as part of a power supply. The band around one end marks the cathode. A group of four such diodes could be packaged into a bridge rectifier.

CHARACTERISTIC CURVE OF A DIODE

The graph in figure 24-9 shows the operating characteristics of a diode. The forward voltage is the voltage that biases the diode 'on' and it conducts. The reverse voltage causes the diode not to conduct or block current. However, if the reverse voltage is made high enough, the diode will breakdown and conduct in the reverse direction.

This (break down) is either the Zener effect if the emf value at breakdown is less than about 5V, or Avalanche effect if it is more than about 5V. This is not a normal operating condition for most semiconductor diodes and may cause lattice damage, ruining the diode.

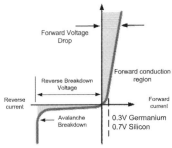

Figure 21-9

ZENER DIODES

The reverse voltage breakdown effect, however, is used in special zener diodes. These diodes are deliberately operated under enough voltage to cause them to conduct in the reverse direction. A resistor must be connected in series with a zener diode to prevent the junction from being destroyed. The circuit across which the diode is

connected will not increase in voltage over the zener breakdown voltage. For this reason, zener diodes are used as shunt (parallel) voltage regulating devices. Zener diodes are used to provide a regulated DC voltage in low power applications. So if some part of a 12 volt DC circuit required 5V DC at low power, then a zener diode with an appropriate series resistor could be used to provide a regulated 5 volt DC output in spite of variations in the 12 volt DC input voltage. A good way to remember the symbol of a zener diode, as shown in figure 24-10, is to note the shape of the cathode line on the diode as representing the forward and reverse current characteristics of a diode shown earlier in figure 24-9.

Figure 21-10

THE CIRCUIT OF A ZENER VOLTAGE REGULATOR

The circuit diagram of a zener used as a voltage regulator is shown in figure 24-11. The unregulated DC to the circuit varies from 8 to 12 volts. Rs is a current limiting resistance to prevent the zener from being destroyed.

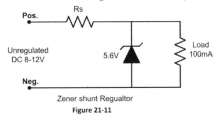

Zener shunt Regualtor
Figure 21-11

Remember, a zener regulator is operated with reverse bias and to the point where it breaks down and conducts in the reverse direction. For some, the term 'breakdown' is confusing. Breakdown does not mean the zener is destroyed or busted! Break down means it is forced to conduct in the opposite direction (from the anode to cathode). Normally a diode operated beyond breakdown is destroyed. However, zener diodes are designed to operate in the breakdown region with a small breakdown current. A zener would be destroyed just like any other diode except for the current limiting resistance Rs. Under these conditions the voltage across the zener is constant. In figure 24-11 the voltage across the zener will be 5.6 volts (it is a 5.6 volt zener - you buy them with a voltage rating). Irrespective of input voltage fluctuations, the voltage across the zener will be a very constant 5.6 volts.

Rs and the zener form a series circuit. The sum of the voltage across the zener (5.6V) and across Rs is equal to the input voltage. Suppose the unregulated input voltage was at a maximum (12V), then the voltage across Rs would be 12-5.6 = 6.4 volts.

As an important rule-of-thumb, the reverse zener current is about 1/10th of the maximum current drawn by the load - the load draws 100mA so one could expect the reverse current of the zener to be 10mA. The zener and the load form a parallel circuit - so the sum of the branch currents is equal to the current through Rs. The current through Rs must then be 100+10=110mA. So we know the maximum current through Rs is 110mA. What is the maximum voltage across Rs? The maximum input voltage is 12V, so the maximum voltage across Rs will be 12-5.6 = 6.4 volts. We can now calculate the resistance of Rs from Ohm's Law:

Rs= E(across Rs)/ I(through Rs)= 6.4 volts/ 110 mA = 58Ω.
Just as important is the power rating of Rs:
Power (of Rs)= E(across Rs) x I(through Rs)= 6.4 volts x 110 mA = 0.704 watt.

EFFICIENCY OF SHUNT REGULATORS

Shunt regulation of voltage is the most inefficient. Look at the circuit of figure 24-11. The current through the zener plus the current through the load is constant. This is because the voltage across the zener and the load is more or less constant at 5.6V. When the load is removed all of the current flows through the zener. So with no load (meaning the regulator is doing nothing for us) maximum current flows through the zener. The zener is then creating maximum heat (losses) when no load is connected.

For this reason shunt regulators are only used in low power applications. For higher power, pass transistors or a darlington pair would be used in conjunction with the shunt regulator. For even higher efficiencies an electronic regulator circuit would be used. For still high efficiencies a Switch Mode Regulator is used. See the discussion of these in the chapter on power supplies.

VARACTOR OR VARICAP

A reverse biased diode (or PN junction) does not conduct. If you refer to when we applied a reverse bias to a PN junction, we saw the width of the depletion zone increased. If we increase the amount of reverse bias further (without reaching breakdown), the width of the depletion zone would widen even more. If we reduce the reverse bias, the depletion zone will narrow. Now, if we continually varied the amount of reverse bias, the width of the depletion zone would also continuously vary. Each time the depletion zone changes in size there must be some movement of electrons in the circuit. However, electrons never move across the junction. The junction, under reverse bias, is an insulator.

Figure 21-12

Does this movement of current on each side of an insulator remind you of capacitance? It should because a reversed biased diode will act just like a small capacitor whose capacitance can be changed by altering the amount of reverse bias. If you like to think of it another way, the depletion zone is the dielectric, which can be made to change in thickness or width by the amount of reverse bias.

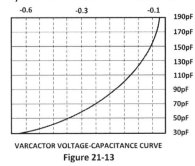

VARCACTOR VOLTAGE-CAPACITANCE CURVE
Figure 21-13

So a reverse biased diode can be used to create a voltage variable capacitor, the symbol of which is shown in figure 24-12. There are purpose made diodes for use as varactors, though in practice almost any diode can be used for this effect. This ability of a diode to behave as a variable capacitance is extremely useful. The frequency of a tuned circuit or a quartz crystal can be made to vary by using a varactor diode. By using a variable resistor to adjust the reverse bias, the capacitance of the diode can be made to vary, which in turn will affect the frequency of the tuned circuit.

LIGHT EMITTING DIODES

In any forward biased diode, free electrons cross the junction and fall into holes. When electrons recombine with holes, they radiate energy. In the rectifier diode, the energy is given off as heat. In the light emitting diode (LED), this energy radiates as light.

It takes energy from the source to move an electron from the valence level to the conduction level. When electrons drop back to the valence level, they will emit this extra energy in the form of photons (light).

Figure 21-14

An electron moving across the PN junction moves to a hole area. This can allow a nearby conduction electron to fall to its valence level, radiating energy. In common diodes and transistors made from germanium, silicon, or gallium arsenide, this electromagnetic radiation is usually at a heat frequency, which is lower than visible light frequencies. With gallium arsenide phosphide the radiation occurs at red light frequencies. Gallium phosphides produce still higher frequency (yellow through to green) radiations. Gallium nitride radiates blue light.

Light emitting diodes come in many colours. The short leg on the diode identifies the cathode so if you want the diode to make light it needs negative on the cathode for forward bias. Typical forward current for the conventional LED is 10-20mA.

There are several photodiodes and photosensitive devices. The photodiodes convert photons to electric emf. There are also a number of other specialist diodes. For examination purposes, we have more than covered enough material here. Also, the depth of the material is more than adequate. You will not be asked to describe the operation of a diode at the electron/hole level, although you should remember the terms used thus far and what they mean.

AN APPLICATION FOR A DIODE

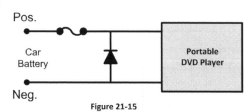

Figure 21-15

Power diodes are made from silicon. Besides being used as rectifiers, one very useful application for a single power diode is reverse voltage protection in radio.

All mobile (used in vehicles) electronic equipment is very much prone to being connected to the source of power the wrong way. I have first hand experience of doing this myself. Connecting the power supply or battery to a DVD player the
wrong way around would be devastating to the DVD player if some sort of reverse polarity protection did not exist. All mobile electronic equipment has an inline fuse. Where the negative and positive leads of the power supply enter the equipment, there is a reverse biased silicon diode. When the power supply polarity is connected the correct way, this diode is reverse biased and acts as an open circuit. If the user accidentally connects the radio to the wrong polarity, the diode becomes forward biased and conducts heavily, blowing the fuse.

Most times when this happens, the diode is destroyed and remains as a short circuit across positive and negative. Shifty technicians will often fix the problem by just cutting one lead of the diode removing it from the circuit and replacing the fuse. Of course, they will charge you too many dollars for this two-minute job! With the diode out of the circuit, the next time the device is connected to reverse polarity you can say goodbye to it for good.

DIODE CLAMPING

A diode will not conduct when forward biased if the amount of forward bias does not exceed the barrier potential. The barrier potential for silicon diodes is approximately 0.6V or 600mV. Suppose we had an application where we had signals that are normally below 600mV but if something was to go wrong and the signal went over 600mV we want some sort of over volt protection.

One such application is a landline telephone receiver. Landline telephones are subject to all sorts of static and other unwanted induced line voltages. The normal operating voltage for a telephone receiver (the bit you put against your ear) is less the 600mV. So if we connect two back to back silicon diodes across the receiver they will have no effect on the receiver's normal operation. Should there be a large static crash or anything else that induces more than 600mV on the telephone line then one or both of the silicon diodes will conduct, effectively short circuiting the telephone earpiece, and saving the user's eardrum.

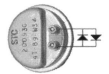

Figure 21-16

This is just one of many applications for diode clamping. It can be used in a receiver or transmitter to clamp a signal path from becoming too high a level. It could be used as the limiter in an FM receiver. Diodes can be configured to clamp either the positive or negative side of a signal. The clamping diodes can be biased to permit a wide range of clamping voltages other than 600mV.

TRANSISTORS

The basic transistor can be thought of as two diode junctions constructed in series. From the bottom, there is a contact against an N-type emitter element. Next to this is a thin P-type base element, with a metal electrode connected to it, forming the first PN junction. A second N-type collector element is added, with a contact on it, forming the second PN junction. This produces an NPN transistor.

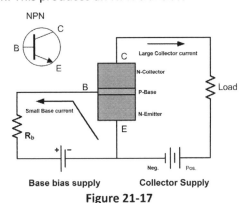

Figure 21-17

This type of transistor is called a bipolar junction transistor (BJT). The important characteristics of construction are:

The base region is very thin and lightly doped.
The emitter region is heavily doped.
The collector is large and usually connected to the case as a heat sink, so that heat can be removed from the transistor.

With just the collector supply connected, no current will flow in the collector circuit as the top PN junction - the one between collector and base - is reverse biased.

Consider when the base-emitter junction is forward biased as shown in figure 24-17. The heavily doped emitter region floods the thin base region with charge carriers (electrons). The base region is lightly doped compared to the emitter. All of the electrons passing across the forward biased base-emitter junction are looking for a hole to fall into. There are more electrons crossing the lower junction than there are holes available on the other side (in the base) to meet them. I like to think of the base region as becoming saturated with charge

179

carriers (which are electrons for an NPN transistor). The excess of electrons come under the influence of the collector voltage (which is higher than the base voltage) and consequently electrons flow in the collector circuit.

The important aspect of the transistor is that *small amounts of base-emitter current can control large amounts of collector current*. If the base current was made to vary, say by the insertion of a carbon microphone (which is a sound dependent resistor), then the collector current will be an amplified representation of the base current. The transistor is an amplifier. Since the BJT transistor uses a small base input current to control a larger output collector current, it is called a current amplifier. Where there is current, there is voltage. Z=E/I and because the current is significant, the BJT transistor is a low input impedance device.

The symbol for a PNP transistor (figure 24-18) is the same as the NPN except the arrow is pointing IN rather than OUT and all the polarities would be reversed.

PNP

Figure 21-18

It is interesting to note that the name transistor comes from "transfer resistor." Another way of looking at the operation is that without the base-emitter junction being forward biased there is no collector current.

When the base-emitter junction is forward biased the resistance between collector and emitter decreases from infinity (or some very high value) and current flows in the emitter circuit. Small variations in the amount of base-emitter current cause the resistance between collector and emitter to vary greatly, but in proportion. Therefore, the collector current faithfully follows the base current but is much larger and supplied with a higher voltage. The collector current is greater than the base current, so the transistor has amplified.

The amount by which an amplifier amplifies is called the gain. The amount by which a transistor amplifies is called Beta and has the Greek symbol β. The beta of a transistor is calculated from:

$$\beta = \frac{\Delta I_c}{\Delta_b}$$

Equation 21-1

The triangle symbol 11 is the Greek letter delta and is the mathematical shorthand for 'change in'. Thus, ΔI_c and ΔI_b are the change in collector and base currents respectively.

So the *Beta (β) or gain of a transistor, is the change in collector current divided by the change in base current.*

A small variation of base current can control 50 to 150 times as much collector current. Thus, the transistor is an ideal control and amplifying device. A junction transistor of this type can be called a bipolar junction transistor (BJT) to differentiate it from a field effect transistor (FET) discussed later.

If a junction transistor has its base-emitter current changed from 10 to 30 milliamps and this causes the collector current to change from 50 to 250 milliamps, what is the current gain or Beta of the transistor? The change in base-emitter current is from 10mA to 30mA = 20 mA. The change in collector current is from 50mA to 250mA = 200 mA. The Beta is therefore 200/20 = 10. The transistor has a current gain of 10.

Comparing a triode and BJT

A triode electron tube is also an amplifier as we have learned. There is one significant difference between a triode and the BJT. Firstly, just to refresh your memory, a triode amplifies by adjusting the negative voltage on the control grid, which in turn can control the large cathode to plate current, resulting in amplification.

A triode is often called a voltage amplifier because it is voltage on the control grid which controls the anode current. Also, because there is no input current to a triode's control grid (Z=E/I), the input impedance of any electron tube is high. A BJT is called a current amplifier because it is base current which controls the much larger collector current. A BJT is a low input impedance amplifier.

The control-grid cathode circuit of a triode does not have any current flowing in it (there are exceptions). This is important. Since a triode can amplify with voltage alone, a triode consumes no power from the input source. From P=EI, if you have no 'I' you have no 'P'. A triode is a high input impedance amplifier, whereas a BJT is a low input impedance amplifier.

Whether a device is a voltage amplifier or current amplifier is irrelevant to the final amplification, though sometimes the input impedance is important. The triode has a significant advantage in being able to amplify a weak signal from a low power source without taking any power from it.

The advantages of the BJT though, are enormous: size, low heat, low power consumption, lower voltages, easier construction and much more.

Amplifier Input Impedance

What is the big deal about input impedance of an amplifier? Take a high input impedance like the electron tube. Only voltage is used on the control grid to control the output current. There is no input current. P=EI; without I there is no P. Suppose an antenna was connected to the control grid. There is only a very small amount of power arriving at the input of a receiver. You cannot cook potatoes with the power

coming off an antenna. The power from an antenna is measured in femto-watts. Femto means 10^{-15}. So an amplifying device that does not need power (high input impedance) is the best to use. If you do take power from an antenna, remembering that it is a resonant circuit, you would drastically lower the Q of the antenna. BJT's are low input impedance; so we have to do tricks with transformers to make them look like a higher input impedance.

Memory Jogger:

Always remember, if you are trying to work out, or asked to work out if a transistor has the correct polarity voltages to operate:

The base-emitter must have forward bias.
The collector-base is reverse biased.
If the bias voltage is correct, current will flow against the arrow in the symbol.

The collector voltage must also permit current flow against the arrow.

A LITTLE ABOUT NOISE

We tend to think of electricity (electron flow) as being fluidic or smooth. This is not correct; electricity is made up of lumps - very small lumps called electrons. Noise is produced whenever an electron does not do what it is supposed to do when it is supposed to do it. In the Electron tube, electrons might collide with secondary electrons emitted from the anode. Imagine a situation in a PN junction where an electron is ready to fall into a hole, but no hole is to be found! For that very small instant that electron represents noise. Any random or unwanted electron motion, or lack of motion, in semiconductor devices, is noise.

Don't get the wrong idea - PN junctions in transistors are wonderful low noise amplifiers, however when it comes to super sensitive receivers like those used for radio astronomy, PN junctions and hole-electron recombination is just too noisy. We shall see later that there are semiconductor devices that amplify and don't have a hole-electron recombination, making them lower noise devices.

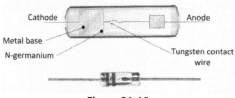

Figure 21-19

POINT CONTACT DIODE

A point-contact germanium (figure 24-19) or silicon diode is a semiconductor pellet (of germanium or silicon) with fine gold-plated tungsten wire with a diameter of about 80 to 400 microns (millionths of a metre) a sharp point, makes contact with the polished top of the semiconductor pellet and is pressed down on it slightly from a spring contact. This cat's whisker, as it is known, is connected on the right hand side to a brass plate which is the cathode. The semiconductor injects electrons into the metal. The energy level between the valence electrons in the semiconductor pellet and the tip of the wire produces a diode action. The contact area exhibits extremely low capacitance.

Because of the low capacitance, point contact diodes can be used for applications at extremely high frequencies. Ordinary PN-junction diodes have too much junction capacitance for use at these high frequencies.

SCHOTTKY DIODE

The Schottky diode (named after the inventor) uses a metal such as gold, silver, or platinum on one side of the junction and doped silicon (usually N-type) on the other side. When the Schottky diode is unbiased, free electrons on the N side are in smaller orbits than the free electrons in the metal. This difference in orbit size is called the Schottky Barrier.

When the diode is forward biased, free electrons on the N side gain enough energy to travel to larger orbits (it takes energy to make an electron move to a larger orbit). Because of this, free electrons can cross the junction and enter the metal, producing a large current. Because the metal has no holes, there is no depletion zone. In an ordinary diode, the depletion zone must be overcome before a diode can conduct - this takes time - a very short time, but time nonetheless. The Schottky diode can switch on and off faster than an ordinary PN junction. In fact, a Schottky diode easily rectifies frequencies above 300MHz. The Schottky is also called the hot-carrier diode. The current carriers are 'hot to trot' as it were and conduction (unlike a conventional diode) takes very little electric pressure (voltage) in the forward direction to conduct. This is an advantage with very weak signals.

The schematic symbol of a Schottky diode:

Figure 21-20

John Bardeen, William Shockley and Walter Brattain in 1948

The first amplifying devices were the electron tubes, They were large in size, consumed a lot of power, produced a lot of heat and were very unreliable. Large equipment using all electron tubes and running all the time required technicians to be on hand at all times just to replace faulty electron tubes.

1930s engineers at American Telephone and Telegraph knew that vacuum-tube circuits would not meet the company's rapidly growing demand for increased phone call capacity. Bell Laboratories' director of research Mervin J. Kelly assigned William Shockley to investigate the possibility of using semiconductor technology to replace tubes.

On December 16, 1947, their research culminated in a successful semiconductor amplifier. Bardeen and Brattain applied two closely-spaced gold contacts held in place by a plastic wedge to the surface of a small slab of high-purity germanium. On December 23 they demonstrated their device to lab officials and in June 1948, Bell Labs publicly announced the revolutionary solid-state device they called a "transistor."

Continuous improvement to the transistor occurred through the 1950s. An important advance in 1954 was the silicon transistor, first from Morris Tanenbaum at Bell Labs and shortly after by a team led by chemist Willis Adcock at Texas Instruments. By the end of the 1950s, silicon had become the industry's preferred material and TI the dominant semiconductor vendor.

in 1959 when, working for Egyptian engineer Martin M. (John) Atalla on the study of semiconductor surfaces at Bell Labs, Korean electrical engineer Dawon Kahng built the first successful field-effect transistor (FET) comprising a sandwich of layers of metal (M - gate), oxide (O - insulation), and silicon (S - semiconductor). The MOSFET, popularly shortened to MOS, promised a significantly smaller, cheaper, and lower power transistor.

22 - Semiconductors II

TRANSISTOR CONFIGURATIONS

There are different ways of connecting a transistor so that it will amplify. The principle of small changes in base current controlling larger changes in collector current remains the same. The configurations we are going to discuss cause the amplifier to have different characteristics.

I am not going to go very deep into this subject, as there is no need to for examination purposes. For the exam, all you really need to be able to do is identify which configuration is illustrated and remember some basic characteristics of each configuration.

The three configurations are:

Common emitter
Common-base and
Common-collector (also called emitter-follower)

I will give you the basic circuits of each. For the exam, you need to identify which circuit is which and that is all. Please do remember that with all transistor amplifier configurations, it is still the small base current controlling the larger collector current which results in amplification. At the end of this section, I will also provide a table of the different characteristics of each configuration. A detailed description is not necessary as this would be rather lengthy and contain much information which is not required.

COMMON-EMITTER CONFIGURATION

The basic transistor amplifier circuit is the common-emitter. It is also the most common configuration you will see and is easy to remember. The circuit shown in figure 25-1 is a basic common-emitter amplifier using an NPN transistor.

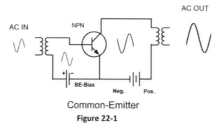

Common-Emitter
Figure 22-1

If a PNP transistor was to be used, the polarity of the batteries would be reversed.

This amplifier circuit uses transformer coupling at both the input and output. The transformers are air cores, so this indicates that the circuit is used at radio frequencies, as iron core transformers have too much loss in the laminated iron core at radio frequencies. The type of coupling does not determine the configuration.

The reason the configuration of figure 25-1 is called common-emitter is because the emitter is the common leg (or lead) between input and output. This is how you will identify the common-emitter circuit.

The identifying feature of this circuit as common-emitter is the emitter leg being common to input and output. One thing worth noting is that the signal being amplified undergoes an 180° phase change. Mutually coupled transformers create a 180° phase change. Notice how the input signal in figure 25-1 goes through a 180° phase change through the input transformer. Common-emitter configuration also causes another 180° phase change as shown at the output of the transistor. Once more the signal passing through the output transformer produces yet again another 180° phase shift.

COMMON-BASE CONFIGURATION

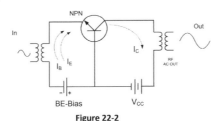

Figure 22-2

The common-base circuit is identified by the common base leg between input and output. In this circuit transformer coupling is once again used.

Transformer coupling in these two circuits is for impedance matching. We have learned that one of the functions of a transformer, if required, is to match two unequal impedances. For example, if you had an audio amplifier with an output impedance of 1000Ω, it would not work well if connected to an 8Ω speaker. An audio output transformer could be used to perform an impedance conversion from 1000Ω to 8Ω.

Just a note, in all the transistor circuits we have seen so far the bias voltage (between base and emitter) and the collector voltage (between collector and emitter) has used separate batteries. In practical circuits, one source of supply is used for base bias and collector voltage. The input signal for both circuits so far is AC. BJT's do not work on AC. However, the base-emitter battery bias not only turns the base-emitter junction on; it converts the incoming AC to VDC (varying DC).

EMITTER-FOLLOWER or COMMON-COLLECTOR

Figure 25-3 shows an emitter-follower, also called a common-collector configuration. I have shown how one source of supply can be used for both base-emitter forward bias and collector voltage. Unlike the previous two circuits, this one is a little different. The key to the identification of the common-collector (emitter-follower) is that the output is taken from across a load or resistor in the emitter circuit. The output of this amplifier and all amplifiers of this configuration is taken from across the emitter load R_L

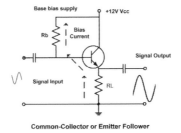

Common-Collector or Emitter Follower

Figure 22-3

185

The only reason the different configurations are used is that the amplifiers provide different characteristics, a summary of which I will provide in table form shortly. They are all just amplifiers, relying on small changes in base current to control larger changes in collector current.

The coupling between the input and output of the circuit in figure 25-3 is called capacitive coupling. Coupling is the method by which signal is fed into and the output is taken away from a transistor amplifier. There will be more on coupling later.

Notice also how a single resistor R_b is used to forward bias the base-emitter junction of the transistor, doing away with having to use two power sources. The resistance of R_b determines the amount of bias current that flows from the negative ground up into the emitter and out of the base, through R_b and back to the positive supply. This bias arrangement is called fixed-bias. I would also like to point out that a number of transistors could be connected via coupling to produce a multistage amplifier.

There you have it. The three basic transistor configurations and an illustration on biasing with a single voltage in the last circuit. We have been using BJTs here. We will be talking about other types of transistors soon and we have already spoken about the electron tube. All of these active devices can be connected in the above configurations.

Summary of the characteristics of each configuration:

Parameter	CE	CB	CC
Input impedance	1000 Ohms	60 Ohms	40,000 Ohms
Output impedance	40,000 Ohms	200,000 Ohms	1000 Ohms
voltage gain	500	800	0.96
Current gain	20	0.95	50
Power gain	10,000	760	48
Phase in/out	180^0	0^0	0^0

Table 22-1

A common-emitter has a low input impedance and a high output impedance and inverts the phase of the incoming signal.

A common-base also has a low input impedance and has a high output impedance but does not create a phase inversion between input and output.

A common-collector has high input impedance and a low output impedance and provides no phase change. A common-collector could drive a low impedance load such as a 50Ω antenna or an 8Ω speaker. The pass-transistors in a low voltage high current power supply are common-collector, as the impedance of most things (like a transceiver) connected to a power supply are low impedance. Think about it 13.8V at 20A for a typical transceiver means a low input impedance R=E/I.

BIASING TRANSISTORS

Biasing transistors is another broad topic that much could be written about. However I am again going to confine this chapter to what I think is the 'need to know' even that may extend further than necessary.

Biasing a transistor means applying a fixed DC voltage between the base and emitter. By varying the amount of bias current, we can make the transistor amplify in the required class of operation i.e. A, B, AB or C.

If you have trouble understanding the need for bias (and many students do) imagine using a transistor without any bias and feeding an AC signal to it to be amplified. Half of the AC signal would not forward bias the base-emitter junction and there would be no output from the transistor during this half cycle. Transistors (including FETs) and electron tubes are not AC devices - they work on varying DC. When we bias the input of an active device at a DC potential, we can then feed an AC signal to that input. The combination of the AC superimposed on top of the DC bias produces varying DC at the input of the transistor or other active device.

We have already seen one method of biasing using a single resistor (R_b) in the emitter follower circuit shown earlier. This type of biasing is called fixed-bias. I suggest you go back and have a look at it now.

Fixed bias is simple. However, it is quite unstable thermally. If the transistor warms for any reason, due to a rise in ambient (surrounding) temperature or due to current flow through it, the collector current increases. The higher the current gain of the transistor the greater the instability of the circuit.

A far superior and the most common biasing arrangement for BJTs, is shown in the schematic circuit of figure 25-4 and is called voltage-divider bias. I have shown typical values for a practical audio amplifier. You will not have to perform any calculations on the circuit, but you may need to identify the circuit, not as a whole but in separate parts so we will have a look at it for the necessary detail.

The common-emitter circuit with voltage divider bias.

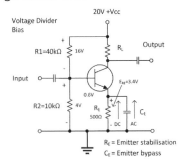

Figure 22-4

The purpose of bias in any BJT (transistor) circuit is to forward bias the base-emitter junction of the transistor. In the circuit of figure 25-4, this is achieved primarily by the series circuit consisting of R1 and R2 in series with the supply voltage (in this case 20 volts) The transistor is an NPN type which requires the base to be positive with respect to the emitter for forward bias. R1 and R2 in series are across the 20 volt supply which is positive at the top (collector) and negative at the earth symbol. Now you could use Ohm's law to approximate the voltage across R1 and R2 knowing they are across a 20 volt supply and have the values shown. Their total resistance is 50kΩ. We get 40/50 (four-fifths) of 20 volts dropped across R1 and 10/50 (one-fifth) of the voltage will be dropped across R2. One fifth of 20 volts is 4 volts. So the approximate voltage across R2 is 4 volts. The top of R2 is positive and the bottom negative. So neglecting R_E, there are about 4 volts applied between the base and the emitter of the transistor, positive at the base and negative at the emitter. The polarity is correct, so we have a forward biased base-emitter junction. This is how voltage divider bias works. R1 and R2 form the voltage divider.

Let's just have a look at some of the other components in the circuit and what they do. Resistor R_E you will notice a relatively low value (500 ohms) and is connected between the emitter and ground. R_E provides thermal stability to the circuit. An increase in DC voltage through R_E caused by a reduction in the transistor collector-emitter resistance will cause an increase in voltage across R_E. The voltage across R_E is in opposition to the bias voltage. If the voltage across R_E was 1 volt, then this opposes the voltage across R2 and so the effective bias voltage is only 3 volts. The voltage across R_E will only increase if the transistor heats up and as a consequence its collector-emitter resistance decreases. Without R_E the transistor could go into thermal runaway.

Let's look at the situation without R_E. The transistor is operating. It heats up due to the current through it or an increase in the ambient (room or surrounding) temperature. If the transistor heats up, its emitter-collector current increases which causes it to heat up more, which causes its resistance to decrease more, which increase the current more, which causes it to heat up more, etc. - this is a thermal runaway and results in the transistor being destroyed.

R_E provides what is called degeneration. Degeneration is a single word, which explains the operation in the second last paragraph. The degeneration used here is also known as inverse current feedback. If the current increases through R_E (due to the transistor heating up for whatever reason) the voltage across R_E reduces the bias voltage, that decreases the emitter-collector current, allowing the transistor to cool and not go into thermal runaway.

Now, it is the DC current flowing through R_E that we are using to provide thermal stabilisation and prevent thermal runaway. There is an AC component in the circuit. This AC component is the signal we are trying to amplify. An AC voltage is applied to the input of the transistor via capacitive coupling and that AC signal will appear in the emitter-collector circuit, amplified. We do not want the AC signal to flow through R_E. If the AC component (the signal to be amplified) did flow through R_E, it would affect the bias of the transistor. So the simple approach is to provide an AC path around R_E for the AC component. This is accomplished by adding the parallel capacitor across R_E (C_E). So, the signal which is being amplified flows through C_E and does not affect the biasing of the transistor, while the DC current, which is what affects the operating characteristic of the transistor and can cause thermal runaway, is allowed to flow through R_E.

If this is difficult to see, let me add the following points to refresh your memory. The circuit of the transistor has a current, this current without any input signal is just DC. When an AC (or varying DC) signal is applied to the transistor, the current through the transistor now has two components (or parts): a DC component, which is responsible for bias; and an AC current which is the signal being amplified. The DC component is used to bias the transistor and prevent thermal runaway - the degenerative voltage across R_E must be treated differently by the transistor than the AC component, which is the signal we are trying to amplify.

If we were to allow the AC component of the current to flow through R_E, then any increase in the signal to be amplified would reduce the gain (amplification) of the transistor, defeating the whole purpose of amplification.

A capacitance will pass an AC current, but block a DC current. So the bias current flows through R_E (DC), while the signal to be amplified passes through C_E. For this to work effectively the capacitive reactance of C_E must, as a rule of thumb, be about 1/10th the value of R_E. So in the circuit of figure 25-4, the X_C of C_E would be about 50 ohms.

Since C_E bypasses the AC component of the emitter-collector current around R_E, C_E is called an emitter bypass capacitor, or just a bypass capacitor. R_E is referred to as the emitter stabilisation resistor.

You do not have to explain how this works as I have done. But you do need to know the purpose of R1, R2, R_E and C_E (which may be labelled differently).

R1 and R2 provide voltage divider bias. R_E provides thermal stability.

C_E bypasses the AC component (i.e. the signal to be amplified around R_E).

RL is just a load resistor. If it were not there, the output would be connected to the +20 DC rail and there would be no output other than +20DC - hardly a useful amplifier.

JUNCTION FIELD EFFECT TRANSISTOR (JFET)

Field effect and bipolar junction transistors are entirely different devices. FETs are solid-state amplifying devices that have operating characteristics very similar to those of triode electron tubes. There are three types of FETs, the junction (JFET) and two metal oxide semiconductor types known as MOSFETs. One is the depletion MOSFET; the other is an enhancement MOSFET.

The essential construction of a JFET is shown in figure 25-5. A block of P-material

called the substrate has an N-type channel running through it. The figure shows a cross section. Think of the N-channel as a wormhole in the P-substrate. As with any PN junction, electrons fall into holes at the barrier and a depletion zone is created. You can see the N-channel running through the P-substrate. The connections are made as shown in figure 25-5. The drain can be thought of as similar to the anode, the source similar to the cathode and the gate similar to the control grid of a triode valve. Though this device is fully manufactured from semiconductors, the purpose of the JFET is to amplify, as is the purpose of the BJT and the triode. If a supply voltage is connected between drain and source, there will be a current flow as there is no PN-junction either forward or reverse biased.

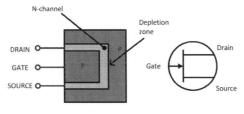

Figure 22-5

So we have connected a supply voltage between drain and source and you can imagine a current flowing from source to drain through the N-channel.

Notice also that the depletion zone occupies part of the channel.

To refresh your memory. A depletion zone is called a depletion zone because it is depleted of charge carriers. A charge carrier is either an electron or a hole. A PN junction can be reverse biased increasing the size of the depletion zone and how much voltage is applied to a PN junction in the reverse bias condition will determine the size of the depletion zone.

If we wanted to reverse bias the JFET in the previous diagram, we would apply a negative potential to the substrate that would attract the holes and a positive potential to the channel that would attract electrons.

If we reverse bias the PN junctions on each side of the channel, we will increase the width of the depletion zone. Since the depletion zone would widen, it would occupy more of the channel. In fact, if we applied enough reverse bias the depletion zone could be made to occupy the entire channel and no current would flow from source to drain. This is called pinch-off.

We have a way of controlling the amount of current that flows in the channel by changing the amount of reverse bias between the gate connection and the channel (the substrate and the gate are connected together).

The gate is always biased negative in the JFET (N-channel), then the signal to be amplified can be applied to the gate. The signal to be amplified would change the reverse bias and hence the size of the depletion zone. Since the depletion zone occupies part of the channel, the resistance of the channel will alter in sympathy with the input signal and in doing so, control the source-drain current.

So the JFET is an amplifier. Since the gate connection is always reverse biased (no current), the input impedance is very high. It is the changes in the negative potential on the gate that controls the width and conductivity of the depletion zone and that in turn controls the much larger drain current.

This is, in essence, exactly how a triode works - by changing the negative potential of the control-grid of a triode we can control the much larger anode current. If the control-grid of a triode is made negative enough, no electrons will flow from the cathode to the anode. This is called cut-off.

Similarly, if the negative potential of the gate of an N-channel JFET is large enough, the depletion zone will occupy the entire channel and there will be no source-drain current. This is called pinch-off.

To run through the operation of an N-channel JEFT one more time, the gate is biased negative with respect to the source, causing the depletion zone to occupy part of the channel. The width of the depletion zone in the channel controls the channel's resistance. If a signal is applied between the gate and the source, it will control proportionately a much larger drain current. The drain current will be an amplified version of the signal applied to the gate.

The JFET is a voltage amplifier like a triode. By this, we mean that changes of voltage only on the gate are required to control the drain current. Since voltage alone is needed to control the JFET, the input signal does not have to deliver any power. You can have all the voltage in the world, but without any current you have no power (P=EI). This is very different to the BJT, which is a current amplifier. You can have voltage without current, but not current without voltage - think about that. The JFET and all FETs are then high input impedance devices.

You could also have a P-channel JFET that would have an N-substrate. The operation would be the same except you would have to bias the gate positive.

Important points to remember: Symbol; arrow in for N-channel; arrow out for P-channel; high input impedance; voltage Amplifier; the semiconductor version of a triode valve; why high input impedance can be an advantage. There are no forward biased PN junctions in a FET; therefore, no electron-hole recombination and therefore less noise.

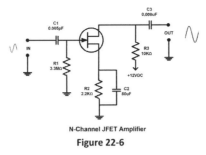

N-Channel JFET Amplifier
Figure 22-6

MOSFETS

Metal Oxide Semiconductor Field Effect Transistors.

When a FET is constructed as shown in figure 25-7 with the gate insulated (silicon dioxide) from the very narrow N-channel, it is called an insulated gate FET, IGFET, or MOSFET, the latter being the most common term. Combinations of these transistors as a single unit are referred to as CMOS, MOS, VMOS.

When a voltage is applied, with negative to the gate and positive to the source, its electrostatic field extends into the N-channel. This reduces its ability to carry current by repelling channel electrons, depleting the channel of its carriers.

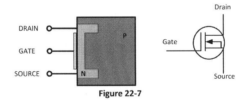

Figure 22-7

This is a depletion mode MOSFET. Sufficient reverse bias can pinch off the drain current completely. Conversely, a forward bias can increase drain current up to a certain point. As a result, this device can operate with no bias at all, which simplifies circuitry. The depletion mode MOSFET is ON without bias. An enhancement mode MOSFET requires a bias to induce a channel and is normally OFF.

The more common enhancement-mode MOSFET or eMOSFET, is the reverse of the depletion-mode type. Here the conducting channel is lightly doped or even undoped making it non-conductive. This results in the device being normally "OFF" (nonconducting) when the gate bias voltage is equal to zero. The circuit symbol shown in figure 25-8 for an enhancement MOSFET transistor uses a broken channel line to signify a normally open non-conducting channel.

The application of a positive gate voltage to an n-type eMOSFET attracts more electrons towards the oxide layer around the gate thereby increasing or enhancing the thickness of the channel allowing more current to flow. This is why this kind of transistor is called an enhancement mode device as the application of a gate voltage enhances the channel.

MOSFETs do not produce any input circuit current due to the insulation between gate and channel. MOSFETs have an extremely high input impedance in the order of 10 megohm. This extremely high input impedance is the MOSFET's great advantage apart from being less noisy than a BJT, this extremely high input impedance means virtually zero drive power. So the source (such as an antenna) does not have to deliver power to a MOSFET to have that signal amplified.

The input impedance of MOSFETs (as much as 40 megohm) and similar devices are so high that they require special handling, as normal handling can result in a static charge being created which can destroy the device. You may find some wrapped in metal foil to protect them from static voltages.

The MOSFET was a significant breakthrough for semiconductor technology and led to combinations of them being fabricated within a single package called an integrated circuit. Of course, even BJTs and JFETs can be fabricated in conjunction with MOSFETs into a single package. It is a simple matter to create resistors in an integrated circuit by controlling the width and doping level of semiconductor strips.

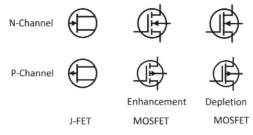

Figure 22-8

Capacitance in an integrated circuit can make use of the fact that a reverse biased PN junction acts as a capacitor. These devices are very high input impedance (2MΩ for the JFET and 10MΩ for the MOSFET). Figure 25-8 shows the symbols for JFETs and MOSFETs for both P and N channel. MOSFETs can have more than one gate. These are called dual-gate MOSFETs. These are ideal for use as mixers where two signals are mixed (with the MOSFET being nonlinear) to produce new frequencies.

PROTECTED GATE MOSFETS

MOSFETs have very high input impedance of 10 Megohms or higher. Because of this high input impedance, even static electricity from handling them can build up a high voltage on the gate (or gates) and rupture the metal oxide insulator. Some MOSFETs are made with internal zener diodes connected in parallel and in opposite directions between the gate and the substrate. Under normal operation, neither of these zeners will conduct. However, if a high static voltage is created on the gate, one of the zeners will conduct and discharge the voltage. These are called protected gate MOSFETs. The symbol is that of an ordinary MOSFET, but within the symbol, you will see one or more zener diodes.

SILICON CONTROLLED RECTIFIERS (SCRs)

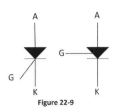

Figure 22-9

The SCR construction shown in figure 25-10 illustrates one of the many NPNP or multilayer semiconductor devices. SCRs are also referred to as Thyristors. The layer structure could also be PNPN. This one is called an SCR. "A" is the anode, "K" is the cathode and "G" the gate. If a voltage, even a large one, is placed across the anode and cathode (of any polarity), this device will not conduct because at least one junction will be reverse biased. The main current path of an SCR is between cathode and anode.

Switch SW1 is normally open (NO). When SW1 is closed, the first P-Type region, acting as the base of an NPN BJT, is forward biased through RG and "load" which allows current to flow through J2. Since J3 is forward biased already, current flows through the whole device from K to A and the load.

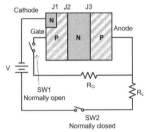

Figure 22-10

The gate switch SW1 can now be opened and current will continue to flow because J2 loses control of its current carriers. Even reverse biasing the gate will not stop the cathode-anode current. It is necessary to reduce the applied voltage (Vaa) to almost zero or open SW2 to stop current flow. No current will flow if the source potential (Vaa) is reversed, even with SW1 closed.

Therefore, with an AC source rather than Vaa, which is DC, the SCR acts like a rectifier in that it allows current to flow in one direction only. The SCR is a unidirectional switch.

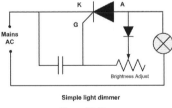

Simple light dimmer

Figure 22-11

An SCR acts like an ordinary rectifier except that it can be switched ON and OFF. When OFF it is an open circuit and does nothing. When ON it acts like any rectifier would. The switching ON can only be done by the gate; the switching OFF can only be done by removing the supply voltage. If the supply voltage is the mains, then on the change of the half cycle on the sinewave the SCR will automatically switch OFF.

SCRs can be made to handle medium to very high currents. A typical application could be a light dimmer or motor speed control circuit (though a TRIAC is better for the latter). The gate of the SCR is not switched (usually) with a mechanical switch instead, some of the mains AC that the SCR is going to control is used to switch or 'fire' the SCR's gate.

An SCR is still a rectifier like any other rectifier, though when it starts to behave as a rectifier depends on the voltage applied to its gate. If you want to control both halves of an AC waveform you could use two SCRs. A device in one package which does this and which has mostly replaced the SCR is called a Triac. Triacs are not in the current (Australian) syllabus.

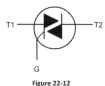

Figure 22-12

Triacs can be considered as back-to-back SCRs and perform as a full wave SCR for alternating current. An SCR is a unidirectional switch and a TRIAC is a bidirectional switch. The gate control voltage of a TRIAC, unlike the SCR, can either be positive or negative in polarity. Figure 25-12 is the schematic symbol of a triac.

OVER-VOLT PROTECTION

One failure mode for many analogue regulated supplies is that the series pass transistor can fail with a short circuit appearing between the collector and emitter. If this happens the full unregulated voltage can appear at the output, and this would place an excess high voltage on the supply output and significant catastrophic damage to the load may occur.

By looking at the voltages involved, it is very easy to see why the inclusion of overvoltage protection is so important. A typical supply may provide 13.8 volts stabilised to logic circuitry. To provide sufficient input voltage to give adequate stabilisation, ripple rejection and the like, the input to the power supply regulator may be in the region of 16 to 20 volts. Even 16 volts would be enough to destroy equipment connected to the load.

SCR/Thyristor over voltage crowbar circuit

The thyristor crowbar circuit shown figure 25-13 is very simple, using only a few components. It can be used within many power supplies and could even be retrofitted in situations where no over-voltage protection may be incorporated.

The SCR over voltage crowbar or protection circuit is connected between the output of the power supply and ground. The zener diode voltage is chosen to be slightly above that of the output rail. Typically, a 13.8V volt rail may run with a 15-16 volt zener diode. When the zener diode voltage is reached, current will flow through the zener and trigger the silicon-controlled rectifier or thyristor. This will then provide a short circuit to ground, thereby protecting the circuitry that is being supplied from any damage and also blowing the fuse that will then remove the voltage from the series regulator. Zener diodes can be connected in series to achieve the 1-2 volts above rail voltage of 13.8V.

As a silicon-controlled rectifier, SCR, or thyristor is able to carry a relatively high current - even quite average devices can conduct five amps and short current peaks of 50 or more amps, cheap devices can provide a very good level of protection for small cost. Also, voltage across the SCR will be low, typically only a volt when it has fired and as a result the heat sinking is not a problem.

The small resistor, often around 100 ohms from the gate of the thyristor or SCR to ground is required so that the zener can supply a reasonable current when it turns on. It also clamps the gate voltage at ground potential until the zener turns on. The capacitor C1 is present to ensure that short spikes do not trigger the circuit. Some optimisation may be required in choosing the correct value although 0.1 microfarads is a good starting point.

If the power supply is to be used with radio transmitters, the filtering on the input to the gate may need to be a little more sophisticated, otherwise RF from the transmitter may get onto the gate and cause false triggering. The capacitor C1 will need to be present, but a small amount of inductance may also help. A ferrite bead may even be sufficient. Experimentation to ensure that the time delay for the thyristor to trigger is not too long against removing the RF. Filtering on the power line to the transmitter can also help.

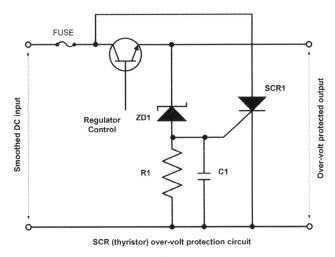

SCR (thyristor) over-volt protection circuit

Figure 22-13

Printed in Great Britain
by Amazon